Darwinism in Philosophy, Social Science and Po..

How much can Darwinian mechanisms account for human values, the character of social institutions and the justification of our claims to knowledge in the sciences? Alexander Rosenberg, the distinguished philosopher of science, explores these questions and their ramifications in this groundbreaking and timely collection of papers.

The essays cover three broad areas related to Darwinian thought and naturalism: the first deals with the solution of philosophical problems such as reductionism, the second with the development of social theories and the third with the intersection of evolutionary biology and economics, political philosophy and public policy. Specific papers deal with naturalistic epistemology, the limits of reductionism, the biological justification of ethics, evolution and the social contract in political philosophy, the political philosophy of biological endowments and the Human Genome Project and its implications for policy.

Many of Rosenberg's important writings on a variety of issues are here organized into a coherent philosophical framework which promises to be a significant and controversial contribution to scholarship in many areas. The book will be of interest to students and professionals in the philosophy of science and the application of evolutionary biology to social science and policy.

Alexander Rosenberg is Professor of Philosophy at the University of Georgia. He received the Lakatos Award for distinguished contribution to philosophy of science in 1993. He is the author of eight other books, including *The Structure of Biological Science* (Cambridge University Press, 1985).

CAMBRIDGE STUDIES IN PHILOSOPHY AND BIOLOGY

General Editor

Michael Ruse   *University of Guelph*

Advisory Board

Michael Donoghue   *Harvard University*
Jean Gayon   *University of Paris*
Jonathan Hodge   *University of Leeds*
Jane Maienschein   *Arizona State University*
Jesús Mosterín   *Instituto de Filosofía* (*Spanish Research Council*)
Elliott Sober   *University of Wisconsin*

Published Titles

Alfred I. Tauber: *The Immune Self: Theory or Metaphor?*
Elliott Sober: *From a Biological Point of View*
Robert Brandon: *Concepts and Methods in Evolutionary Biology*
Peter Godfrey-Smith: *Complexity and the Function of Mind in Nature*
William A. Rottschaefer: *The Biology and Psychology of Moral Agency*
Sahotra Sarkar: *Genetics and Reductionism*
Jean Gayon: *Darwinism's Struggle for Survival*
Jane Maienschein and Michael Ruse (eds.): *Biology and the Foundation of Ethics*
Richard Creath and Jane Maienschein (eds.): *Epistemology and Biology*
Jack Wilson: *Biological Individuality*

Darwinism in Philosophy, Social Science and Policy

ALEXANDER ROSENBERG
University of Georgia

CAMBRIDGE
UNIVERSITY PRESS

PUBLISHED BY THE PRESS SYNDICATE OF THE UNIVERSITY OF CAMBRIDGE
The Pitt Building, Trumpington Street, Cambridge, United Kingdom

CAMBRIDGE UNIVERSITY PRESS
The Edinburgh Building, Cambridge CB2 2RU, UK   http://www.cup.cam.ac.uk
40 West 20th Street, New York, NY 10011-4211, USA   http://www.cup.org
10 Stamford Road, Oakleigh, Melbourne 3166, Australia
Ruiz de Alarcón 13, 28014 Madrid, Spain

First published 2000

Printed in the United States of America

Typeface Times Roman 10.25/13 pt.   *System* LATEX 2ε   [TB]

A catalog record for this book is available from the British Library.

Library of Congress Cataloging in Publication Data
Rosenberg, Alexander, 1946–
Darwinism in philosophy, social science and policy / Alexander
Rosenberg.
p.   cm. – (Cambridge studies in philosophy and biology)
Includes bibliographical references and index.
ISBN 0-521-66297-4
1. Naturalism.   2. Social Darwinism.   I. Title.   II. Series.
B828.2.R66   2000
146 – dc21   99–38044
CIP

ISBN 0 521 66297 4 hardback
ISBN 0 521 66407 1 paperback

For
LARRY WRIGHT
who made Darwinism possible
and in memory of
MARTIN HOLLIS
who made it difficult

Contents

Introduction

Following the model of *The Origin of Species*, the books I have written are each one long argument. The essays collected together and slightly revised here have not figured in any of these long arguments. But they do reflect a common theme: the implications of a broader naturalism and a more specific Darwinism for issues with which nonbiologists concern themselves. The oldest of them goes back only about ten years, and in each case I still find myself happy with their conclusions, if not altogether any longer satisfied with their expression and argument. Though written by a committed naturalist and one of Darwin's latter-day "bull dogs," these papers give voice to the recognition that there are important limits to the power of these two inspirations to solve problems in philosophy, science and policy.

What is meant by 'naturalism' is something the first of these essays, "A Field Guide to Recent Species of Naturalism," more fully recounts. But briefly, naturalism in latter-day philosophy is founded on a commitment, voiced initially by W. V. O. Quine, to let the sciences be our guides in epistemology and metaphysics. The renaissance in evolutionary biology and philosophers' increased recognition of its relevance to human affairs have made a generation of naturalists into defenders and exponents of the theory of natural selection as naturalism's most informative guide. This is a conviction which I share. But I also recognize that naturalism leaves hostages to philosophical fortune: problems of justification that cannot be ignored. Detailing the structure of naturalism in the philosophy of science and how it deals with outstanding questions, especially of realism and antirealism, enables us to gauge its strengths and weaknesses.

The second essay in this section, "Naturalistic Epistemology for Eliminative Materialists," is an attempt to press the case for a Darwinian-inspired naturalism about a central philosophical project: the establishment of an epistemology. With the towering exception of Quine, philosophers have despaired

1

of a thoroughgoing naturalistic epistemology. This is largely because they have insisted that epistemology must be normative, and most philosophers hold that naturalism has no normative resources. This latter claim is one I treat at greater length in the essays in section three. The former claim is controverted in the present essay, which draws upon the resources of a naturalistic epistemology that Quine could, indeed should, endorse.

If, as naturalism holds, a biological theory is to be our guide in philosophy, or the human sciences, we need to understand its character, methods, scope and limits. Along with many other philosophers of biology I have sought to understand and express biology's scope and limits. The next three essays in this volume reflect a fairly well settled conclusion that biology is an instrumental discipline, one whose conceptual form is shaped more heavily by our cognitive limitations and practical interests than are the physical sciences so shaped. "Limits to Biological Knowledge" argues that the nature of biological theory and explanation reflects a compromise between a search for nomological universality bequeathed by success in physics and the operation of variation and selection in the biocosm we inhabit. In this essay I explain in a clearer way, I hope, than in previous books and papers, exactly how evolution inhibits generality in our best biological theories.

The relentless change dictated by selection over variation in changing environments and the biologist's search for laws combine to narrow the prospects for reduction of less general to more general theory in biology and reconfigure reductionism as a philosophical thesis as well. How nature and our search for causal theories conspire to stimulate and frustrate reductionism is the subject of the other two essays in this section. In "Reductionism Redux: Computing the Embryo," I attempt to reconcile the limits of reductionism with the rapid expansion in molecular biology and especially our understanding of the molecular biology of development. The essay reports the extent to which pessimism about our understanding of the molecular mechanisms of embryological development has been overtaken by events in late-20th-century biology. If anything, the discoveries of molecular developmental biology have fostered confidence among biological reductionists even as philosophical ones have been reducing their claims. "Reductionism Redux" seeks to find an intellectually satisfactory compromise between the reductionist biologists and the antireductionism now widespread even among the most tough-minded physicalist philosophers of science. I argue that the physicalist antireductionism consensus of contemporary philosophy of biology is tenable only on an instrumentalist treatment of biological theories and explanations.

The third essay in this section, "What Happens to Genetics When Holism Runs Amok?" provides a new and different argument for my instrumental perspective on biology, one which starts by accepting the premises of "developmental systems theory," a reconceptualization of divisions in biological theory that reflects a path of convergent evolution in the thinking of biologists and philosophers towards the view I develop in the second essay of this collection. Developmental Systems Theory argues against a separate science of genetics on the ground that the distinctions it requires between genetic signal and environmental channel, and between organism and ecosystem, do not divide nature at the joints. I agree with the philosophical arguments developmental systems theorists advance; indeed, I think they can be strengthened, and I attempt to do so. But, as the essay argues, these purely philosophical arguments leave actual biology and its explanatory theory pretty much as it was. This raises again, in sharper focus, the fundamental question posed in "Limits to Biological Knowledge" of just what biological understanding consists in.

Earlier in the twentieth century naturalism also included a commitment to grounding substantive normative judgments of morality and political philosophy on scientific foundations. In the last two decades the explanatory power for human affairs of evolutionary biology and the theories it has inspired have once again made ethical naturalism attractive. Taking naturalism this far is a temptation to which I have not succumbed, and the next three chapters in this collection explain why. Very broadly my objections to grounding morality on science are those of David Hume and G. E. Moore: you cannot validly infer a statement about what ought to be the case from one that describes merely what is the case, no matter how innocuous the former and how far-sighted the latter. "The Biological Justification of Ethics: A Best-Case Scenario" attempts to explain why ethical rabbits cannot be pulled from evolutionary top hats. It goes on to the somewhat more favorable prospects for explaining our ethical principles and political arrangements by appeal to their adaptational consequences for individuals in the evolutionary line of descent from which we trace our origins. The attempt to found morality on an understanding of evolutionary processes often exploits findings and theories in naturalistic, evolutionarily inspired social science as well as primatology. "Moral Realism and Social Science" argues that the addition of resources from these disciplines is not likely to enhance the prospects for a non-normative foundation for the rightness of ethical principles or political institutions. But again, we may go some way towards explaining how we ended up sharing the same moral intuitions and broadly similar institutions across cultures by attending to their adaptational consequences in an environment such as ours and our ancestors'. "Contractaranism and the 'Trolley' Problem" illustrates these prospects by exploiting one of the most salient of "science fiction" cases with which moral philosophers have wrestled in the last two decades, the so-called trolley problem.

The last set of chapters reflects the intersection of my interests in biology with another long-standing preoccupation: economics and policy. Along with, indeed prior to, the onslaught of biology on my intellectual attention in the 1970s, problems at the intersection of philosophy and economic theory have long preoccupied me. As my study of evolutionary biology progressed, however, I came to see important similarities between its explanatory strategy and that of economics. In books and papers over the last fifteen years or so I have reflected on the parallel in both theories' commitment to maximization – in the one case of fitness and the other of utility – and how this commitment is reflected in the sharing of mathematical tools and equilibria as explanatory concepts by both disciplines.

Others, notably economists, have also sought to make methodological common cause between their discipline and evolutionary biology. These attempts have resulted in no serious improvement on the vexed state of development in economic theory. This should be no surprise, given the powerful misunderstanding of the theory of natural selection evinced in these economists' writings and the differences, less visible but more significant than the similarities, between economic theory and evolutionary theory. The first essay in this last section, "Does Evolutionary Theory Give Comfort or Inspiration to Economics?" identifies the attractions and shows the crucial differences between evolutionary theory and economics that make neoclassical economics a scientifically frustrating enterprise, and evolutionary economics a nonstarter.

But applying biology – molecular and evolutionary – to other important issues is not so fraught. As the last two essays show, a little knowledge need not always be a dangerous thing. At least, what little I know of molecular, evolutionary and functional biology is deployed, along with some economics, in "The Political Philosophy of Biological Endowments: Some Considerations" to help us deal with a fundamental problem of welfare economics: Assume that one aim of government is to ensure some sort of equality of outcome and not merely a procedural equality of opportunity; or again, if equality of outcome is too strong a goal, surely attaining real equality of opportunity even for those with disabilities is a policy goal on which almost all will concur. The question then arises how we are to take account of the unequal and unearned distribution of talents and abilities, weaknesses and disabilities among individuals we hope to set equal in opportunity (or outcome). This policy problem has vexed a number of political philosophers. In this essay I bring together biology and economics to shed light upon some popular solutions to the problem.

Finally, in "Research Tactics and Economic Strategies: The Case of the Human Genome Project," I explore the interaction between fundamental research in nucleic acids and the broadest questions of contemporary science policy, both public and private. The question at issue is the scientific rationale for the human genome project, and the non-scientific incentives that molecular biologists have to advocate it as a priority project for national and international governmental support.

A few of these essays have been revised since their first publication as a result of discussion and criticism. But because their conclusions remain settled in my own mind as roughly correct, collecting them together here provides a fairly accurate reflection both of what I consider important areas of research to explore further and the general approach with which I hope to continue to explore them.

Since the conclusions several of the essays argue for are negative, suggesting limitations on biology and other endeavors inspired by it, one might well ask what I suppose is worth continued exploration. The answer is not hard to provide. If the essays in part two of this collection are on the right track, philosophers of biology – whether employed in philosophy departments or biology departments – have much to do simply to uncover the nature of understanding and explanation as they operate in biology. The bland assurances of physicalist antireductionism can no more survive scrutiny in the philosophy of biology than in the philosophy of psychology. We need to replace it, by either an instrumental view of biology or a new account of the nature of biological explanation.

As for the biological basis of ethics – the non-normative task of explaining how our moral intuitions and the codes which they reflect came about – there is much more to be said, and much being said, by game theorists, anthropologists and evolutionary psychologists, as well as philosophers. Moreover, no one can pretend that either Hume or G. E. Moore has said the last word about inferences from what is or is not the case to what it ought to be. So, those of us who remain convinced that no ethical teaching can be constructed from any purely factual foundations must continue to defend this claim.

Finally, just as no one could have predicted the coming of the Human Genome Project, no one could have anticipated the policy issues it raised. There is no doubt that without even counting the questions of policy and political philosophy which genetic testing, germ-line manipulation, laissez-faire eugenics and cloning raise, the continuing revolution in molecular biology will raise successive intellectual problems for philosophy. Within the limitations of naturalism and Darwinism, there remains more intellectual excitement and potential than almost anywhere else on the philosophical and scientific horizon.

1

A Field Guide to Recent Species of Naturalism

In his presidential address to the American Philosophical Association in 1955 Ernest Nagel expounded and defended a philosophy he called *naturalism*, despite the fact that "the number of distinguishable doctrines for which the word 'naturalism' has been a counter in the history of thought is notorious . . . " [Nagel, 1956, p. 3]. In retrospect what was distinctive about Nagel's naturalism was its radical departure from the modern western tradition in epistemology. Nagel rejected as an objection to naturalism the charge that it does not secure independently any warrant for the epistemic role it accords the methods of science. According to this objection, in committing itself to the logic of scientific proof without further foundations, naturalism is quite analogous to religious belief in resting on unsupported and indemonstrable faith. Such an argument, Nagel wrote,

> has force only for those whose ideas of reason is demonstration and therefore refuse to dignify anything as genuine knowledge unless it is demonstrated from self luminous and self evident premises. But if, as I also think, that ideal is not universally appropriate, and if furthermore, a wholesale justification for knowledge and its method is an unreasonable demand and a misplaced effort, the objection appears quite pointless.
>
> The warrant for a proposition does not derive from a faith in the uniformity of nature, or any other principle with cosmic scope. The warrant derives exclusively from the specific evidence available, and from the contingent historical fact that the special ways employed in obtaining and appraising the evidence have generally been effective in yielding reliable knowledge. [p. 15]

This latter principle is the Archimedian point of naturalism. Upon it, the lever of latter-day naturalism has moved much of the debate in moral philosophy, the philosophy of mind, epistemology and the philosophy of science. In all these areas naturalists have advanced a series of distinctive theses all of which

Nagel, and for that matter John Dewey, would recognize as familiar and friendly amendments to their philosophy.

Interestingly enough the philosophers who have come so to style themselves owe as much to the pragmatism of W. V. O. Quine as they do to Nagel's tradition. For it was "Two Dogmas of Empiricism" [in Quine, 1951] which more than any other work made philosophical space for naturalism as a viable alternative to positivism and ordinary language philosophy.

Once the distinction between the analytic and the synthetic is surrendered, the demarcation between scientific theory and philosophical analysis is extirpated. Philosophy neither needs to nor can it defend itself as merely the logical clarification of thought. It need no longer fear to tread the ground hitherto reserved for science, and it must be assessed on the same standards as very abstract scientific theory. For that is the only thing philosophy can be for those who accepted "Two Dogmas." But 'naturalism' has come to label that movement in philosophy which is willing to adopt Quine's arguments against empiricism, but not to go the whole way with Quine towards eliminativism about modality or meaning. This is what makes naturalists the protegés of Ernest Nagel. "Epistemology Naturalized" [in Quine, 1969] is not the locus classicus of philosophical naturalism. Rather, it is *The Structure of Science* [Nagel, 1961].

Naturalists in most of the subdisciplines of philosophy have viewed their project as one that seeks a reasonable compromise between competing extremes while doing as much justice to the extreme views as they deserve. Thus, in the philosophy of psychology, naturalists seek to steer a course between the Scylla of dualism and the Charybdis of eliminativism. So too, in the philosophy of science more broadly, naturalism seeks to reconcile the history of science as a human institution with the positivist insight that its methods are uniquely suited to providing objective knowledge certifiably independent of its social and psychological context. By doing so, naturalism hopes to provide a viable alternative to the constructivist relativism which swept through the social studies of science in the 1980s and the "constructive empiricism" [van Fraassen, 1979] of philosophers who despair of a less radical defense of the objectivity of science.

Though Nagel expressed willingness to exploit scientific theory to underwrite philosophical conclusions, he did not actually do so. His naturalistic successors have shown no such reluctance. The appeal to the details of scientific theory has characterized naturalism in several compartments of philosophy.

Naturalism in moral philosophy has rejected the arguments of Hume and Moore that normative claims cannot follow from descriptive ones. Under the label "Moral Realism" ethical naturalists [for example, Railton, 1986] have appealed to empirical theory about the evolution of social organization to underpin an account of the moral good. In epistemology, under the name "reliablism," naturalists have claimed that it is for science to uncover those non-deviant causal chains on which we can rely to provide truth a high percentage of the time. Such causal lore will justify our beliefs in a way that frees them from Gettier-type problems. Philosophers of psychology from Dennett [see Dennett, 1969; Millikan, 1984] onwards have adopted the label 'naturalism' to express their commitment to a project for reconciling intentional psychology with materialism through the good offices of the theory of natural selection.

Darwinism has a special role for latter-day naturalism. In this respect naturalism has transformed itself in a way Nagel might not recognize. Invoking Isaiah Berlin's simile of the fox and the hedgehog, Nagel wrote "We have come to think . . . that there are a great many things which are already known or remain to be discovered, but that there is no one 'big thing' which, if known, would make everything else coherent, and unlock the mystery of creation. . . . We have come to direct our best energies at the resolution of limited problems . . . that emerge in the analysis of scientific discourse, in the interpretation and validation of ethical and aesthetic judgments, and in the assessment of types of human experience." Nagel rejected "system building." But many contemporary naturalists have become hedgehogs. They have concluded that there is one big thing that makes almost everything else coherent. They share a Darwinian approach to philosophical theory so thoroughly, that it would be easy to synthesize their views into a traditional philosophical system.[1]

The fascination that Darwinism has come to have for latter-day naturalists is easy to explain. Most of the concepts for which philosophers have struggled to propound a naturalistic account are functional and therefore teleological. Certainly all the intentional concepts are, as Taylor established [1964]. Therefore most of the epistemic concepts are teleological too, as are normative notions that characterize actions, their consequences, and the rules which govern them. Now, Nagel thought that teleology was to be naturalized by appeal to directively organized systems [see Nagel, 1961, chapter 11]. His successors have found the etiological account originally propounded by Wright [1976] more convincing. And the relevant etiology is natural selection over blind variation.

A second reason for the fascination that Darwin exercises in philosophy reflects advances in evolutionary biology which have dramatically increased its claim to explain aspects of human affairs. Of all the well-confirmed theories in modern science, it is the one with most direct relevance for the human condition, human behavior, and its cognitive causes. If any well-established scientific theory can teach us about ourselves it is Darwin's. Other theories, which might teach us more, which might even limit the writ of Darwinian theory for understanding human affairs, are either so far not well-confirmed, or even well-formed. If naturalism is to replace a priori first philosophy with scientific theory, then at least for the nonce the theory in question will be Darwin's.

So, we may characterize naturalism in philosophy as follows:

1. The repudiation of "first philosophy." Epistemology is not to be treated as a propaedeutic to the acquisition of further knowledge.
2. Scientism. The sciences – from physics to psychology and even occasionally sociology, their methods and findings – are to be the guide to epistemology and metaphysics. But the more well established the finding and method the greater the reliance philosophy may place upon it. And physics embodies the most well established methods and findings.
3. Darwinism. To a large extent Darwinian theory is to be both the model of scientific theorizing and the guide to philosophical theory, because it maximally combines relevance to human affairs and well-foundedness.

Naturalism in the philosophy of science adds another element to this credo:

4. Progressivity. Arguments from the history or sociology of science to the non-rationality, or non-cumulativity, or non-progressive character of science, are all either unsound and/or invalid.

Even though for most naturalists in the philosophy of science this last provision is a theorem derived from one or more of the first four principles, it is important enough to secure pride of place among them.

1 REALISM'S NEED FOR NATURALISM

In a famous passage of *The Structure of Science* Nagel damned realism and instrumentalism about science with fine impartiality:

> It is ... difficult to escape the conclusion that when the two apparently opposing views on the cognitive status of theories are each stated with some circumspection, each can assimilate not only the ... subject matter explored by experimental inquiry but all the relevant facts concerning the logic ... of science. In brief the opposition between these views is a conflict over preferred modes of speech. [Nagel, 1961, p. 152]

Nagel dismissed the dispute between realism and instrumentalism on the verificationist ground that neither side made a testable difference for our account of the nature of science. Once positivist and post-positivist injunctions against metaphysical questions lapsed, philosophers of science returned to the vexed question of realism and anti-realism[2] about the cognitive status of scientific theories. It was realists – those who hold that our scientific theories not only embody truths about unobservables but ones we can know – who early recognized that their view required naturalism.

Richard Boyd has been a persistent realist almost since the debate between realism and instrumentalism returned to center stage in philosophy of science in the early seventies. One of his most sustained published defenses of realism, "The Current Status of Scientific Realism" [in Leplin, 1984], devotes as much attention to a refutation of what Boyd calls "constructivism" as it does to a refutation of empiricist agnosticism about scientific theory. Neither sociological constructivism nor van Fraassen's empiricism can explain what Boyd calls the instrumental success of scientific theorizing. They cannot answer the question "why should so theory-dependent a methodology be reliable at producing knowledge about (largely theory-independent) observable phenomena" [Leplin, 1984, p. 57]. Only realism – the thesis that our scientific theories are approximately true, and increasing in their approximation to the truth – can explain the instrumental reliability of science. But this argument faces a challenge that the abductive principle it relies on – the conclusion of an argument to the best explanation is well-justified – requires independent warrant.

It is at this point that the realist calls upon epistemic naturalism: "Like the 'naturalistic' epistemologists, the scientific realist must deny that the most basic principles of inductive inference or justification are defensible a priori. In a word, the scientific realist must see epistemology as an *empirical* science" [Leplin, p. 65; see also Boyd, 1980, pp. 625–627]. The epistemic naturalist's causal theory of perception also appeals to an explanatory abduction from sensory experiences to their distal causes. But, Boyd notes, "the causal theorist's position does not stand or fall on the strength of that abduction in isolation. Instead the alternative empirical and naturalist conceptions of knowledge, and especially of the epistemic role of the senses, must be evaluated as rival philosophical theories" [p. 75]. And philosophical theories are scientific ones [p. 65].[3]

2 NATURALISM WITHOUT REALISM

It did not take long for anti-realist philosophers both to reject the data from which Boydian realism infers to its best explanation and the principle of inference he employed to reach realism as an explanatory theory. What is more, some of realism's keenest opponents insisted that their anti-realism had at least as much, indeed more right to the title of naturalism, unreasonably arrogated by realists.

Most effective of these naturalistic opponents, or at least the one given the most respectful audience in the philosophy of science, is Larry Laudan. Because he has been a highly effective critic of relativist, constructivist approaches to science, which deny its objectivity [see for instance Laudan, 1981a], and because of his adherence to naturalism, Laudan shares a great deal with realists. But this has not prevented him from subjecting the doctrine's view of the history of science to a withering attack. If Laudan is right, then naturalism does not require realism. But deprived of realism's devotion to truth as the attainable aim of theory, naturalism does need an account of epistemic value which Laudan declines to provide.

Laudan accepts that the sciences show increasing instrumental reliability. But, the explanans to which scientific realism appeals – the increasingly approximate truth of science – could not explain science's reliability. For the history of science is not the history of theories successively approximating more and more closely to the truth, limning more and more completely the furniture of the universe. Laudan's "pessimistic induction" focuses on a number of the most important cases of scientific advance, and argues that advances though they undeniably were, the objects over which they quantified, and the kinds they deemed natural, were entirely superseded by subsequent scientific theories. Laudan instances Ptolemy's physical cosmology, phlogiston theory, and Fresnel's ether theory, each of which had significant empirical warrant in novel prediction, and constituted important advances in their respective disciplines. None of them, however, can be said to have been reverentially successful in the way realism requires [Laudan, 1981b, 1994a]. By induction, we should attach no more confidence to the referential success of contemporary scientific theory as a metaphysical disclosure of the ultimate character of reality. The instrumental success of science cannot be explained as a consequence of its successive approximation to the truth, because the evidence does not unequivocally suggest that it is enhancing its approximation to the truth.

This conclusion shifts the burden of explaining the instrumental success of science on to Laudan's shoulders. It is here that naturalism, and in particular recourse to principle 3 above, enters: we cannot appeal to *a priori* considerations to explain science's past successes, underwrite its future prospects, and justify its methodology. We need to appeal to scientific factors:

> Although we appraise methodological rules by asking whether they conduce to cognitive ends, . . . the factors that settle the question are often drawn from . . .

11

the level of factual inquiry. Factual information comes to play a role in the assessment of methodological claims precisely because we are continuously learning new things about the world and about ourselves as observers of the world.

Laudan provides an illustration:

contrary to popular conception, the superiority of randomly collected data over non-randomly is not a discovery made by mathematics or formal logic as a feature of inquiry in all possible worlds. It is because the world we inhabit turns out to be so uncooperative, we find it appropriate to insist on a variety of stringent rules about sampling in order to achieve our cognitive ends. [Laudan, 1984, p. 38]

According to Laudan, we should conceive methodological principles as resting on claims about the empirical world, claims to be assessed in precisely the same way in which we test empirical theories. We do so by recasting methodological prescriptions as conditionals about the effective means to stated ends. Our most well established substantive theories help us determine whether these recast conditionals are well supported. If they are not, we may revise either the conditionals' antecedent in order more efficiently to attain our epistemic aim, or surrender the aim as unobtainable and substitute another one. This is very clearly a naturalistic view, one which turns scientific findings into epistemological considerations.

Since "science has been successful at producing the epistemic goods" [Laudan, 1987, p. 28], methodology needs to attend to its findings. Laudan agrees that science promotes ends we find cognitively important, and has become progressively more successful over time. But, in light of his repudiation of the inference from the instrumental success of science to its approximate truth, what sort of epistemic goods can Laudan admit science delivers, and how can they inform methodology? If science isn't progressing in the direction of the truth, in what direction is it progressing? Laudan's answers to these two questions are disarming: In *Science and Values* [1987] he declines to identify some central and constant aim of inquiry which science is increasingly successful in meeting. This is because the aims of science vary over time, discipline and even researcher. Laudan does not quite tell us that there is no such thing as a transcendent aim of inquiry, to which all other goals are instrumental means. But he insists that a study of the actual aims and actual methods of scientists undercuts claims the realist makes about what these aims are. The naturalism in Laudan's philosophy comes in the identification of science's and scientists' actual transitory aims as the right ones for science.

To explain the success of science, Laudan distinguishes theories, methods and aims of scientists who employ the methods and embrace the theories.

12

According to Laudan's "reticulated model" theories, aims and methods interact to adjust to one another in a continuing cycle of revisions that insure
the persistent attainment of scientific goals. For a philosopher of science, it
is obvious how method can influence theory choice; for a naturalist, it should
be clear how theory can influence method. Laudan's study of the history
of science reveals the extent to which the aims of science and of scientists
have changed over time, and he shows how changes in these aims result from
changes in the theories and methods scientists embrace and employ. For example, the post-Newtonian methodological proscription of unobservable entities
gave way in the early 19th century to a method of hypotheses that encouraged
them; it did so because of the striking success of theories in chemistry and
electricity which appealed to theoretical entities. During the first half of the
19th century both theory and methodology together undermined the cognitive
aims identified in a strict empiricist epistemology. The aims of inquiry thus
changed from empirical adequacy to truth. And these new aims and methods of course encouraged further theoretical development [Laudan, 1984,
pp. 55–60]. We might add that from the late 19th century onward, the empirical and conceptual obstacles that atomism faced again shifted methodology
away from hypotheses about unobservables and back towards the epistemic
aim of empirical adequacy.

However, Laudan is no historical relativist. Along with other naturalists
he holds that it is permissible to assess prior theories and methods in the light
of our current cognitive goals. We can and must draw conclusions about the
progressiveness of methodological and theoretical change relative to us. And
the reticulated model shows we can use our knowledge of available methods of inquiry as a tool for assessing viability of proposed cognitive aims
as well [p. 68]. Laudan recognizes that "all this sounds rather 'whiggish',
and so it should be, for when we ask whether science has progressed we are
typically asking whether the diachronic development of science has furthered
cognitive ends that we deem to be worthy or desirable" [1984, p. 65]. Laudan
does not tell us what these ends are. Laudan holds that "to lay down a set of
cognitive aims . . . would undermine much of [my] analysis. For . . . the aims
of science vary, and quite appropriately so, from one epoch to another, from
one scientific field to another, and sometimes among researchers in the same
field" [1984, p. 138]. Nevertheless, we may safely conclude that whatever
the current goals of science, a) in the light of Laudan's pessimistic induction, there is no reason to suppose current goals will always be with us; and
b) these goals do not include the successive approximation to the truth, because
if they did so, Boyd's type of realism would be vindicated; but c) the goals of
science, whatever they are, underwrite the claims of science to "deliver the

goods," and undermine the suggestion of relativists and constructivists that science has no epistemic goals or a plurality of incommensurable ones.

But in the end, Laudan's doctrine cannot retain a dignified silence about the goals of science, nor can it condemn all epistemic goals as transient. To do so would open the field to relativist views of sociologists of science which Laudan considers far more mischievous than short-sighted, overzealous scientific realism, or even narrow-minded constructive empiricism. If Laudan's philosophy of science really is a form of naturalism, if it endorses thesis 4 above about the long-term cumulativity of science, then Laudan owes us a naturalistic account of some permanent and enduring (set of) scientific goal(s). In "Progress or Rationality? The Prospects for Normative Naturalism," Laudan recognizes that "a theory of scientific progress needs an axiology of inquiry, whose function is to certify or de-certify certain proposed aims as legitimate" [Laudan, 1987, p. 29]. This need, felt especially keenly by naturalism, is one to which we will return with increased urgency below.

3   NATURALISM IS IN THE DETAILS

By the late eighties, scientism – the central thesis of naturalism – was coming increasingly to be embraced by philosophers of science as the dialectically most powerful move in their now lively controversy with sociologists and historians of science eager to discredit science's right to cognitive "hegemony." By embracing the thesis that we must look to science for the justification of science's own methods, these philosophers showed their respect for "the actual practice of science," while nevertheless invoking the findings and methods of their preferred sciences and scientists. Much of this epistemologically relevant science was and remains fallible. But the tentative character of this philosophically relevant science was not allowed to qualify the strength of philosophers' convictions about the nature of the sciences very much.

Giere's *Explaining Science: A Cognitive Approach* [Giere, 1988] takes its subtitle seriously enough to invoke some of the tentative theories of cognitive psychology in the service of naturalism. Giere's book buys into all three of the axioms and the one theorem of naturalism identified above. The sciences neither have nor need distinctively philosophical foundations, in large part because there is no distinctively philosophical theory, "[T]here is only deep theory, which is part of science itself" [p. xvi]. But Giere goes a little further than other naturalists. Not content with assimilating philosophy to science, he leaves no central role for philosophers in developing this "deep theory." "The people best equipped to engage in such pursuits are not those trained as philosophers, but those totally immersed in the scientific subject matter – namely scientists" [p. xvi]. The effect of this dictum on Giere is to accept physicists' commitment to realism, and to reconcile it with philosophical theories and theses to which physicists are quite indifferent. Thus Giere seeks an account of science that will bring naturalism back from Laudan's agnosticism about successive approximation to the truth: "There is room for a modest yet robust scientific realism that insists scientists are at least sometimes successful in their attempts to represent the causal structure of the world" [p. xvi].

Giere's realism is tempered by his apparent acceptance of the historical record as Laudan reads it – the pessimistic induction. "No realistic theory of science can be viable if it fails to account for the historical evidence Laudan presents" [p. 46]. Ironically, the way realism can deal with the historical record is by adopting an account of theorizing developed first and most fully in the interests of anti-realism: Van Fraassen's semantic approach, according to which a theory is not an axiomatic, syntactic hypothetico-deductive system, but a sequence of models. Van Fraassen's anti-realist motivation for this approach may have been to find an account of theories in which truth – approximate or otherwise – need not figure even metaphysically as a property – known or unknown – of theories.[4] Giere embraces the semantic approach however owing to the empirical claim that scientists themselves seek to develop and present their theoretical results in the form of models. A theory is a set of models, which are themselves idealized abstract formulae and their interpretations, along with various hypotheses linking these models with systems in the real world.

The model-theoretic approach secures its claim to realism by substituting "similarity" for (approximate) truth as a property successive models bear to some designated real system. The similarity between models and real systems in the world is of course partial. Nevertheless, Giere tells us that "the notion of *similarity* between models and real systems provides a much needed resource for understanding approximation in science. For one thing it eliminates the need for a bastard semantical relationship – approximate truth . . . for another it . . . reveals . . . that approximation has at least *two* dimensions: approximation in respects and approximation in degrees. Armed with just these distinctions, we can begin to attack . . . objections to realism" [p. 106]. Thus Laudan's pessimistic induction from the failure of the key terms of successful theories to refer to the unqualified falsity of these theories is short-circuited:

[Laudan's argument against realism] rests on the unstated assumption that approximation is always a matter of *degrees*. If the ether does not exist, claims involving the ether can't be just a little bit off. They must be mistaken in some more radical sense. The argument collapses, however, if we abandon talk of approximate truth in favor of similarity between the model and the world which allows approximation to include respects as well as degrees of similarity.

> Whether the ether exists or not, there are many respects in which electro-magnetic radiation is like a disturbance in an ether. Ether theories are thus . . . approximations. The fact that there is no ether . . . [is a good basis for] reject-ing ether models, but not for denying *all* realistically understood claims about similarities between ether models and the world. [Giere, 1988, p. 109]

There is a well-known problem with similarity-to-real-systems as a dimension along which models may improve in realism. As Wittgenstein is purported to have argued, anything can be similar to anything else, depending on the features compared. Similarity, it may be concluded, is a social construction, and not an objective relation between models and nature.

More important, judgments of overall increases in similarity require the comparability of qualitatively different features of models, and the aggre-gation of the differing "respects" along which similarity is measured. What Giere's argument needs at a minimum is a naturalistic theory of similarity, just as Laudan needs a naturalistic axiology for cognitive inquiry. Giere's nat-uralism is reflected in his appeal to Darwinian theory to provide this account of similarity:

> The effect of evolution on our sensory apparatus is known to have been par-ticularly strong. Animals are capable of incredibly fine discriminations among objects in their environment without the benefit of social conventions. . . . For at least some perceptual judgments, . . . the fact of widespread agreement does not require a social explanation. The explanations of evolutionary biology and physiology are sufficient. [p. 109]

Presumably what these biological explanations are sufficient for is identifying the objective, unconstructed dimensions along which similarity of models to reality is to be measured.

But this appeal is clearly inadequate to Giere's needs. It is by no means clear that the discriminations, that we have been selected for making, give the dimensions of similarity-to-reality on which realism about theory can be established. At most, Darwinian considerations will underwrite successive models as increasing in local survival value, not in some absolute similarity to reality. Suppose that the order in which perceptual and cognitive discrim-inations are salient to survival gives the order in which successive models are graded for similarity to real systems. Then we will inevitably identify as more realistic among competing theories ones which reflect those discrimina-tions our environment has already selected us to make in the past. It will not

necessarily make real distinctions that obtain independently of these discrim-
inations. Here Giere's naturalism goes uncritical. We will see another naturalist, Kitcher, succumb to the same Darwinian temptation in the next section.[5]

Giere notes that like rats and wasps, humans have been selected for being
able to construct and retain representations of the way the world is. Rats and
wasps "have the capacity to construct internal maps of their environment in
any way that yields useful similarities with that world" [p. 110]. The philosophy of science should come to grips with the similar powers of *Homo sapiens*,
by taking note of recent work in cognitive psychology, especially the work of
Kahneman and Tversky [1982] and Herbert Simon [1957]. The former pair
show us that scientists are not "Bayesian agents," while Simon's "satisficing
account of how scientists choose one model over another meets the goal of
being both 'naturalistic' and 'realistic'. It appeals only to natural decision
strategies and it permits hypotheses to be understood realistically" [p. 168].

Despite the effort to weave a bit of substantive cognitive psychology into
his philosophy of science, Giere's fundamental argument for realism within
the ambit of naturalism turns out to be a variant of Boyd's inference to the
best explanation. This time it is an inference to the best explanation of what
scientists do, instead of what they achieve: For Giere, "the only remotely plausible *scientific* account of what . . . physicists are doing requires us to invoke
entities with roughly the properties physicists ascribe to protons and neu-
trons" [p. 111]. And what physicists are doing is manipulating and controlling
[pp. 125–126]. ". . . If this claim is correct, realism provides the proper account
for at least some important areas of modern science" [p. 111]. The emphasis
on a *scientific account* is in the original, and it reflects the priority of naturalism over realism. The philosophical thesis is underwritten by a scientific
claim. Thus, Giere's argument for realism follows from his commitment to
naturalism, just as Boyd's does. Yet, as we have seen, these realists share a
prior commitment to naturalism with at least one anti-realist – Laudan. Indeed, Laudan can embrace almost all of the picture of science Giere offers,
without swallowing the realism at all. Laudan can do so, provided he offers
an account of epistemic goals to replace Giere's increases in respects and
degrees of similarity. Such an account will however have to honor naturalism
about epistemic aims in the way Giere's appeal to Darwin attempts to.

4 NATURALISM FOR ALL SEASONS

For the centennial issue of the *Philosophical Review* [Kitcher, 1992], Philip
Kitcher was invited to discuss the recent history of epistemology. The title

he chose was "The Naturalist Returns," and the paper endorses all of the theses identified above in the even more generalized versions needed to sustain naturalism in epistemology. And yet the word 'naturalism' hardly figures in Kitcher's magnum opus, *The Advancement of Science* [Oxford, 1993]. Like Hamlet's father in Elsinore, however, its presence is everywhere felt. For Kitcher's aim is nothing less than a magisterial reconciliation of every reasonable opinion under the canopy of an all encompassing naturalistic establishment of the objectivity of science. It is a compromise which credits almost every participant in the enterprise over the last two score years with a valuable contribution that will need further exploration.

Kitcher sets himself the prescriptive task of giving an account of the goals of science, and deriving from it what counts as progress towards this goal [p. 61]. In this respect he is less reticent than Laudan. But as a good naturalist Kitcher recognizes that goals are constrained by actual means: ought implies can, and can't implies need not. So, like Giere, he offers us a sketch of some recent cognitive science designed to show what the scientist qua sentient creature can do by way of information storage, and retrieval. Kitcher's embrace of a propositional theory of cognition (contra eliminativists like Churchland) is tentative, except for a commitment to the "thesis that perception provides a theory-independent basis, intersubjectively available, for checking and correction of theoretical claims" [p. 66].

Like psychology, the history and actual character of the sciences plays a large role for Kitcher too. Darwin's achievement, and Lavoisier's, in particular, determine the shape that Kitcher's philosophy of science accords to theories. Where Giere's reflection on textbook presentations of Newtonian mechanics leads him to accord central place to models, Kitcher's absorption with biology in particular makes the notions of an "explanatory pattern" and "consensus practice" central. In biology there is little to mirror the logical positivist's expectation of laws of increasing generality deductively systematizing empirical generalizations. So, scientific theories cannot be like this. Instead, Kitcher holds, theories are answers to *significant* questions. The structure of these answers are general argument patterns: sequences of "schematic sentences," together with instructions for turning them into real sentences about a variety of subject matters that instantiate the schemata, and a classification of the schematic sentences into basic and derived [p. 82]. Examples of theories Kitcher sketches include answers to significant questions about compound weights given by the successive schemata of Daltonian chemistry, its 19th century refinement as valence chemistry, and its 20th century development into electron shell-filling.

A "consensus practice" is the set of explanatory patterns accepted within a scientific community, along with the lines of authority and the division of labor that the development of these explanatory patterns and their interrelations suggest. These consensus practices are the units of scientific change. Progress is a matter of successive consensus practices more nearly attaining the goals of science.

About these goals Kitcher is not reticent. Recall the notion of a *significant* question in individuating theories. The goal of science is not truth, it is significant truths, or rather, increasingly verisimilar answers to significant questions. To understand what makes a question significant requires a little more apparatus. Explanatory schemata always have problematical presuppositions – assumptions the community has some reason not to accept. There are also instances in which their application seems problematical. A question is significant if its answer either a) shows how a correct schema comes to be instantiated – it reveals underlying mechanisms, or b) shows how the schema can be instantiated in problematical instances, or c) shows how one of its problematical presuppositions can be instantiated, or d) the answer is instrumental to answering a question whose answers do one of a, b or c.

But the erotetic progress such answers would provide is far too abstract, and too weak to meet Kitcher's self-imposed need for "goals for the project of inquiry that all people share – or ought to share" [p. 92]. The most radical relativist can accept the typology of explanatory schema, consensus practice, and advance in answering questions significant in the four-fold way, while denying the naturalist's claim that science progresses *absolutely*. So Kitcher's real claim is that the erotetic progress of science is achieved by articulating schemata that identify and employ natural kinds and reveal objective dependencies among their instances [p. 95]. Questions are significant in the end if their answers help us identify these kinds and dependencies. An older way of expressing these goals of inquiry is the discovery of projectable predicates and the reduction of the laws in which they figure. So why not dispense with "significant questions" and simply hold that theories are schemata that uncover these kinds and dependencies? The answer is that objectivity is too complex a notion to be forced into the procrustean bed of Ernest Nagel's structure of science. Too little of the science Kitcher and other naturalists consider objective – especially evolutionary biology – fits nicely into this bed.

But, one may ask, why are uncovering kinds and dependencies the goals of inquiry? Consistency demands that the answer must be a naturalistic one. Like Boyd and Giere, Kitcher recognizes that realism could underwrite uncovering natural kinds and objective dependencies as epistemic goals successively approached by explanatory patterns [p. 96n]. But explanatory improvement might also consist in producing schemata that more fully meet some other non-realist criterion of adequacy, for example, the criterion of increased explanatory unification. Elsewhere Kitcher has opted for an avowedly non-realist unification as the goal of scientific explanation. In *The Advancement of Science* he declines to defend this view, evidently holding or hoping that naturalism's argument to the objectivity of science need not take sides for or against realism:

> The treatment of many questions about progress can profitably be undertaken by leaving the notions of natural kind and objective dependency temporarily unanalyzed, using them to explicate conceptions of progress and only returning to the metaphysics later. So this chapter [and indeed the entire volume] will remain neutral between the options of strong realism and the type of weak (Kantian) realism that I have previously espoused. [p. 96n]

The weak (Kantian) realism in question explicitly absorbs claims about the causal structure of the world into claims about the logical relations among explanatory schema [Kitcher, 1989, p. 477]:

> ...the 'because' of causation is always derivative from the 'because' of explanation. In learning to talk about causes or counterfactuals we are absorbing earlier generations' views of the structure of nature, where those views arise from attempts to achieve a unified account of the phenomena.

In other words "weak (Kantian) realism" is fairly robust transcendental idealism.

Despite his qualms, however, the remainder of *The Advancement of Science* provides a legion of arguments that defend realism against its strongest naturalistic critic, Laudan.

First Kitcher needs to secure the intelligibility of the notion of verisimilitude employed to mark progress in identifying natural kinds and objective dependencies. He does this simply enough by limiting his need for increasing approximation to truth to significant individual statements about the values of meter readings at instants, not general statements about causal relations everywhere and always. He claims not to require an account of increasing approximation to truth for generalizations: "success in achieving exceptionless generalizations is by no means a sine qua non for good science.... Darwinian evolutionary biology has served us as an example" [p. 121]. It is easy to see how predictions about observations improve, as the number of decimal places in which agreement between predicted and observed measurements increases. But, according to Kitcher, almost everywhere else progress is not a matter of successive approximation to the truth. This is where *significance* comes in: "We can loosely sum up [the] advances [of Fresnel's wave theory over Brewster's corpuscular theory] by suggesting that . . . Fresnel was closer to the truth than Brewster – but this is only shorthand for a complex of relations" [p. 123]: relations between the details of their practices which enable us to see that Fresnel made conceptual and explanatory advances, was able to answer significant questions about interference and diffraction that Brewster could not, in addition to making magnitude ascriptions closer to the observed values than Brewster was able to. According to Kitcher successive approximation to the truth is only one component of a much more complicated sort of progress along many different dimensions. And realism – the doctrine that science shows such approximation – now really is watered down, or enriched, by the qualifications, conditions and elaborations that *The Advancement of Science* prescribes.

Despite the qualifications, Kitcher really does feel the need to vindicate the realist claim that science increasingly approximates to significant truths against Laudan's pessimistic induction. He does so by challenging Laudan's examples of successful theories that fail to secure reference for their explanatory variables, and so cannot be true, even approximately. Kitcher's "general diagnosis of what goes wrong in the examples on Laudan's list" is that "[e]ither the analysis is not sufficiently fine grained to see that the sources of error are not involved in the apparent successes of past science or there are flawed views about reference; in some instances both errors combine" [p. 143]. The focus of Kitcher's diagnosis is Laudan's crucial claim that the key concept, the 'ether' of Fresnel's strikingly successful theory, fails to refer. Kitcher argues that this term does not in fact figure directly in the schemata Fresnel employed to successfully answer significant questions about diffraction, interference, polarization, the bright spot at the center of the shadow of a circular disk, etc. The sole function of the ether in most of Fresnel's writings is to provide a medium of propagation.

The account of ether's role in Fresnel's theory leads Kitcher to a distinction between "working posits" – like wave – the putative referents of terms that occur in problem-solving schemata; and "presuppositional posits" – like ether – entities that presumably must exist if instances of the schema are to be true. Kitcher's claim is that as science progresses, presuppositional posits are surrendered (as a new explanatory pattern replaces a problematical presupposition in an older one), but working posits remain, because they do the work of answering significant questions. Thus, Kitcher's naturalist, like Boyd's, draws the conclusion of an inference to the best explanation of why working posits work: they successfully secure reference. Whence the vindication of realism about science's ability to limn the ontology of nature.

Though he does not use Boyd's name for the argument, Kitcher sets out to vindicate inference-to-the-best-explanation against the challenge of van Fraassen's attack on realism. The vindication is vintage naturalism in action. Van Fraassen famously holds that the goal of science should be empirical adequacy – predictive power with respect to observations: science cannot aim at truths – significant or otherwise – about matters unobservable. Challenged to answer Boyd's question, van Fraassen replies that the instrumentally reliable theories are those which have survived Darwinian selection.

Taking this simile seriously, Kitcher patiently lays out the features of Darwinian inquiry: first, enumerating current and past species that have endured; second, attributing comparative fitness levels on the basis of species endurance; third, explaining the fitness by identifying generic fitness-making characteristics; and fourth, describing in detail the way organisms achieve these characteristics. Only a superficial Darwinism would settle for the first two stages in this inquiry. *Mutatis mutandis* for scientific theories: philosophy of science should enumerate current and past theories, note which theories have in fact been most widely espoused because of their explanatory and predictive success, then seek the generic characteristics that endow theories with explanatory and predictive power, and the particular ways in which theories achieve these generic characteristics. Van Fraassen's stopping point – explaining survival of theories by their predictive power – is comparable to the superficial Darwinism that explains survival by appeal to fitness without in turn explaining fitness in any detail.

Note how the force of Kitcher's argument relies on the primacy of scientific theory to guide philosophical argument. Van Fraassen would be better advised to decline to answer the question of why science has succeeded, on the ground that no answer to it would enhance the empirical adequacy of science.

5   NATURALISM AND KNOWLEDGE

But how can we be confident that science – whether interpreted realistically or otherwise – is objective knowledge of significant truths, even loosely interpreted? What we need is an epistemology. In sketching Kitcher's account of epistemic justification – one more fully elaborated than those of other naturalists – we need to keep in mind the matter of whether this epistemology has a naturalistic foundation. For if it does not, then there will be some deeper commitment in Kitcher's philosophy of science than the four theses identified as the core of naturalism when we began.

For Kitcher the logic of inductive justification relies on the elimination of alternatives. The received explanatory schemata of consensus practices restricts the alternative possibilities to a finite number. All but one rival are then eliminated by evidence that inquiry generates. Eliminativism makes the basic relation of confirmation comparative: confirmation obtains between evidence and two or more hypotheses, one of which is most well confirmed among them. Treating the basic confirmatory relation as comparative enables Kitcher to circumvent the classical paradoxes generated by the notion of a positive instance for theories of confirmation as a two-place relation between one hypothesis and the evidence. But what are we to make of ties in the confirmational contest, of cases in which the evidence cannot choose among hypotheses, cases of underdetermination?

The short answer Kitcher gives is that it doesn't happen: "the notion that theories are inevitably underdetermined by experience, is a product of the under-representation of scientific practice" [p. 247]. Deploying rich detail from his account of schemata and practices, Kitcher shows that "Once the constraining character of prior practice is understood, there is no quick argument for the widespread presence of underdetermination on Quinean [Duhemian] grounds" [p. 251]. Much harder to mitigate is the argument for underdetermination inspired by Kuhn. This argument begins by noting that the costs of various adjustments in a theory to preserve it from falsification will vary. And the differences among them will be a function of the significance accorded to the questions answered by the theoretical components to be adjusted. If, as Kuhn held, differing scientists measure the costs of adjustments (to continue the metaphor) in non-convertible currencies, it will be impossible to adjudicate resulting alternative explanatory schemata and incommensurable consensus practices may proliferate. Underdetermination thus leads to a well-known form of relativism.

Kitcher responds that such a "rash judgment . . . can be tempered by reminding ourselves that there are instances in which preferring one response requires dismissing the significance of a large number of predictions and explanations and taking debate to revolve solely around one difficulty" [p. 252]. He braces this response with studies calculated to show that classical paradigm shifts reveal sufficient evidential bases to undercut any suggestion that they might have been underdetermined and theoretically question begging. Kitcher details the process whereby an explanatory schema is revised in response to inconsistency with evidence to show how constraints on revisions usually preclude underdetermination, even through substantial changes in consensus practice.

Is Kitcher's inductive eliminativism a description of how science proceeds or a prescription for how it ought to proceed? Since it is offered as a version of naturalism, eliminativism must be both description and prescription. Indeed, it must secure its prescriptive force for methodology and epistemology on its descriptive adequacy for human cognition as revealed by science. Kitcher admits as much when he addresses the chicken-and-egg question of which came first, the hypotheses to which we apply eliminative induction or the method itself. If the former, then some hypotheses cannot have been justified by eliminative induction. If the method antedates hypotheses about the world, on what basis did it arise and how came it to be justified?

> ... the eliminative propensity is overlaid on a more primitive propensity to generalize. As a consequence of our genotype and our early developmental environments, human beings come initially to categorize the world in a particular way, to view certain kinds of things as dependent on others, to generalize from single instances of especially salient types. Moreover, ... there is a propensity to restrict those generalizations in particular ways when matters go awry.... [T]his *primitive apparatus* works tolerably well in confronting the problems our hominid ancestors encountered; it is relatively well designed for enabling primates with certain capacities and limitations to cope with savannah environment and with the complexities of a primate society.... [W]ith those practices in place, the eliminative propensity can be activated, ... and allows for the ... revision of the primitive categories and views of dependence.... [W]ith these practices in place, the eliminative propensity can be activated, and the use of that propensity ... allows for modification of practice, the revision of primitive categorizations and views of dependence. [p. 241]

Here we find the crux of Kitcher's naturalism, a barely acknowledged commitment to the notion that *Homo sapiens* have been selected for starting cognitive inquiry on the right track. That is, primitive humans began by categorizing and hierarchizing nature in ways that submitted of refinement towards the goals of science by eliminativist methods. Among the infinitude of alternative initial positions of inquiry, we started out from one that included the realist's right answers to significant questions as an attainable outcome.

It is at this point that naturalistic anti-realism launches its counterattack. To begin with, Laudan will not leave Kitcher the high ground of history. If Fresnel's commitment to ether was not a working posit, but only a presuppositional one, Laudan will find a successful theory in which the working presupposition fails to refer. One case so developed is Ptolemy's postulate of crystalline spheres, which enabled him to calculate the order and distances of the moon, sun and planets in close agreement with empirical data [Laudan,

1994a]. Here a working posit has no referent whatsoever, on even the most generous construal.

But history will not settle this debate. It lacks a metric to measure predictive success and successful reference, and an agreed catalog of test cases. So, Laudan challenges Kitcher's move from eliminativist induction to realism. Like Kitcher, Laudan accepts that the fundamental notion of confirmation is relational, or "comparative": e confirms H′ more than H″. But, Laudan says, comparative confirmation is too weak to meet the realist's needs, since realism holds out the hope that theories are improving on an absolute metric, not merely a comparative one. A hypothesis may be most well confirmed among known alternatives and yet be far from well-confirmed, still less approximately true. Without an independent reason to suppose that the available alternative hypotheses have high probability, the identification of the one among them with the highest probability can do little to substantiate a realist's claim that it is approximately true, or that its working posits refer, or that the questions it answers are the really significant ones.

Laudan's point is not that realism is complacent in its endorsement of the received world picture. He does not for a moment suppose that science ought to begin actively seeking more alternative hypotheses beyond those which already provide a good account for the known phenomena. On the contrary, he charges that realism plus eliminativism requires science to engage in wild goose chases in order to find all the hypotheses needed to convert "H′ is more well confirmed than H‴" into "H′ is well-confirmed." Kitcher's realist metaphysics – its endorsement of science – is inconsistent with his comparativist epistemology.

What is more, Laudan adds to his pessimist induction from successful theories whose variables failed to secure reference, an optimistic induction. He identifies theories superseded but which made a comeback, once novel auxiliary hypotheses not contemplated by the superseding explanatory schemata suddenly became available. Thus, to take a striking and debatable example, the 19th century catastrophists' theory of large-scale species extinction via a deluge has been revived in the last 20 years by Alvarez's evidence of cometary impact and dinosaur extinction. This case and others are supposed to show that it is hard to eliminate every known alternative once and for all. Thus optimistic induction from past theories casts further doubt on the progressiveness of science as a purely comparative enterprise in which cases of underdetermination are few and far between.

None of these problems tax naturalism sans realism, suggests Laudan. Comparativism, eliminativism without a commitment to approaching transcendent goals like truth and absolute significance, is untroubled by the existence of unformulated rivals to actual explanatory schemata. It need not fear the possibility that novel auxiliary hypotheses may recall failed theories to new explanatory service. Agnosticism serves better in the naturalists' campaign against the relativist followers of Kuhn. While a realist must show the improbability of an unformulated theory equally well confirmed by evidence as is actual theory, the non-realist naturalist need only insist that reasonable judgments be made among known alternative theories. Anti-realism shifts the burden of proof onto the relativist to identify an alternative theory as equally well confirmed as the latest teachings of real science.

6 NATURALISM'S DARWINIAN ROOTS

Who is the real naturalist? The realist or the anti-realist? Can both be naturalists? Does naturalism have to take a stand between them? The answer becomes clear when we consider how Kitcher might respond to Laudan's arguments for naturalism without metaphysics, and without an agreed-upon axiology.

Recall the observation made at the outset of this chapter of the centrality of Darwinian theory to naturalism, not only in philosophy of science, but in epistemology, philosophy of mind, and for that matter naturalistic moral philosophy and metaethics. Kitcher's own appeals to Darwinian theory as historical exemplar and as methodological mentor are persistent.[6] Recall that the refutation of van Fraassen's constructive empiricism is Darwinian. More important, the assurance that the initial point of inquiry on which eliminativism has worked to refine science in the direction of correct answers to really significant questions, is evolutionary as well.

The passage was quoted at length in section 5 above: we were, as one might say, "preselected" for confirmational eliminativism and for success in science. Kitcher may respond to Laudan that the realist eliminativist need not spend eternity seeking all the hitherto unformulated alternatives. The realist needn't have all the possible hypotheses before beginning to eliminate in the order of absolutely (as opposed to comparatively) less well confirmed. Because the theory of natural selection, the scientific theory which most fully combines confirmation and relevance to human cognition, provides all the assurance realism needs: our cognitive adaptation guarantees that the starting point is one from which the realist's objectives for science will be attained, world enough and time permitting.[7]

Notice this is thoroughgoing naturalism: it underwrites a philosophical thesis – realism – by appeal to a scientific theory, and not just any theory, but evolutionary theory – the naturalist's favorite scientific theory.

One problem with this argument is that it overreaches the science it is based upon. There is as yet not enough in our evolutionary biology to suggest that selective forces shaped our perceptual and cognitive systems to recognize the real natural kinds and the actual objective dependencies in nature. Indeed, if anything, behavioral biology may have shown the contrary. Thus, recent work following Edwin Land suggests that Locke was right about colors being secondary qualities, and not natural kinds [Hardin, 1992]. And other work by Toomey and Cosmides [1993], for example, suggests that the sort of reasoning processes characteristic of science were selected not for their ability to enable us to uncover objective dependencies, but to detect reciprocators and cheaters in contexts of social cooperation. Perhaps, the environment of the savannah selected for a cognitive system that classifies in the way we have come to do so not because the classification is correct, but because it is the only compromise between competing exigencies like predator/prey discrimination, calculating and reasoning speed versus accuracy, complete memory store versus random access with loss. For all we know, we are genetically endowed with categories in rather the way Kant supposed. But not because they are *a priori* and necessary, but because they are quick and dirty adaptations.

Even if our conceptual scheme is designed to start from a point of inquiry close to the truth, evolutionary biology is decades or centuries away from being able to provide a detailed warrant of the sort Kitcher needs for naturalism to go realist. For, by the standards of Kitcher's own critique of van Fraassen, it is only a superficial version of Darwinism that infers from the existence of our explanatory schemata to their fitness as means to the goals of science.

But Kitcher must be granted an appeal to Darwin here, on Laudan's own grounds. For among extant hypotheses, it is the most well confirmed, and the burden of proof may fairly be placed on the anti-realist to identify an alternative adaptational or non-adaptational explanation for the apparent agreement about kinds and dependencies characteristic of prescientific peoples, and the rapidity with which they come to embrace science's kinds as natural and its dependencies as objective.

But the intellectual coherence of naturalism faces a graver threat than ungrounded confirmational eliminativism. This is the charge that naturalism must ultimately be a question-begging or viciously circular philosophy of science. The charge is serious because of the role of vicious circle arguments in naturalism's own critique of opposing views in epistemology, the philosophy of psychology, and elsewhere.

Though naturalists may disagree among themselves about what the goals are that science successfully secures or increasingly approaches, naturalism argues for the progressiveness of science. For this reason we may rely on present science to illuminate and explain the progressiveness of past science. And we may rely on it to provide reasonable constraints on future scientific theorizing in the form of non-trivial methodological prescriptions and the exclusion of a large number of conceivable research programs as unproductive. More fundamentally, we may read off the philosophical principles that describe the nature, extent and justification of scientific knowledge from successful science itself. Does this idea, scientism, proposition 3 among the fundamental principles noted at the outset, the cornerstone of naturalism, require independent justification? It does need justification, if and only if circular reasoning – albeit reasoning in a large circle – lacks probative force.[8]

Recall, that latter-day naturalism begins with Quine's repudiation of the analytic/synthetic distinction, on the ground that all the ways of introducing it are *circular*. Recall, again, that naturalism in the philosophy of psychology is at great pains to "discharge" homunculi – to cash in intentionality for non-intentionality because psychological explanations that fail to do so degenerate into a *regress* or a *vicious circle*. It cannot avail the naturalist to argue that in the philosophy of science a principle honored by naturalism elsewhere has no force.

One alternative response to the vicious circle argument that naturalists cannot afford to contemplate is an appeal to the greater coherence of their package of principles than the package of either constructive empiricism or relativism-*cum*-constructivm about science, nor by a *tu quoque*.[9]

Naturalism cannot tolerate virtuous circles, illuminating infinite regresses or the selection of alternative theories on the grounds of coherence alone, because natural science and natural scientists do not! It would be patently inconsistency to extol science as the source of methodological prescriptions and then employ methods proscribed by these very prescriptions. And that includes science's proscription of inconsistency!

Naturalism needs to break out of the circle of arguments it advances, not just to refute alternatives, but to establish its own internal coherence. There is only one resource it can count on to do so. All the naturalists we have canvassed agree that science, in Laudan's terms, delivers the goods. What are the goods science delivers? Perhaps significant truths, but at least truths about what we can expect to experience. Naturalism has no truck with the denial of an observation language that can frame statements about middle-sized objects in our vicinity. It holds that science enables us to anticipate and to exploit expectations about such items; naturalism holds that even when disappointed expectations underdetermine theory, current science is better at predicting and controlling them than past science, and science gets better at it every day. Among naturalists, both realists and anti-realists vindicate current scientific theory as knowledge because of its impressive record of prediction and control.

Advancing prediction and control as the aims of inquiry is perhaps the only way naturalism can hope to transcend mere coherence in its claims for science as objective knowledge. The strategy involves comparing naturalism's claim about the goals of inquiry with its chief competitors: constructive empiricism and historical relativism-*cum*-social constructivism.

Constructive empiricism differs from naturalism in that it sets out prediction and control as the goals of science, but only as a second best, not as the goals of inquiry *überhaupt*. For constructive empiricism sets out an *a priori* philosophical theory – empiricism – about the aims of inquiry – truth – which it tells us theoretical science cannot in principle attain. Accordingly empiricism condemns science to seek something less than significant truth, by methods underwritten *a priori*: the goal is empirical adequacy and the methods are those consistent with empiricism. For this view the final court of epistemic appeal is an *a priori* theory, and the aim of inquiry is truth.

Relativism-*cum*-constructivism is a form of sophistry, in a historical sense: it holds, with Protagoras, that man, or rather social groups, are the measure of all things: that the aim of inquiry is to convince, persuade, to gain acceptance by whatever means are effective. The methods of science are thus ultimately underwritten by politically driven ideology. The erection of prediction and control as the aims of inquiry would simply reflect the temporary hegemony of one particular group, say the classical physicists, as opposed to another group of scientists, for example the evolutionary biologists. Relativism-*cum*-constructivism rejects the question of what the goals of inquiry ought to be, in favor of the question what do they happen to be for the nonce. And it does so because it denies the very possibility of epistemology. For this view the final court of appeal is group popularity, and the aim of inquiry is to secure its approbation.

Naturalism makes prediction and control the goals of inquiry, and it concludes that well-confirmed science – all of it, not just the empirically adequate parts – is knowledge because it attains these goals. Naturalism identifies those as the goals of inquiry not because an *a priori* philosophical theory so identifies them. This would be patently inconsistent for naturalism, since it denies there is an *a priori* first philosophy. Naturalism certainly does not identify prediction and control as the goals of inquiry in light of any majority vote of currently practicing scientists. The pessimistic induction alone is enough to temper enthusiasm for what scientists happen to embrace for the moment. For naturalism the final court of epistemic appeal is scientific method, because scientific theory attains the aim of inquiry: predictive control.

It is tempting to take another step down this path: to appeal to science to underwrite prediction and control's status as the ultimate aims of inquiry. But, as we have seen, this argument would be circular in a way naturalism will not abide: Science is underwritten as objective knowledge because of its predictive success; predictive success is the criterion of objective knowledge because science – especially Darwinian theory – tells us we require it to survive.

But without another step, how does naturalism press the claims of prediction as the enduring aim of inquiry against the empiricist's claim for truth, and the constructivist's claim for nothing? Does the philosophy of science and epistemology end in a cul-de-sac of incommensurable values?

Naturalism might seek recourse to some transcendental argument for predictive power as a mark of knowledge: an argument which purports to show that any adequate epistemology must make long-term increases in predictive power a necessary condition for new knowledge. But short of begging the question against its competitors, it is hard to see how such an argument might be effected.

One option that naturalism cannot assay is that offered by pragmatism – the philosophy that makes successful prediction and control sufficient and not just necessary for knowledge. Pragmatism is a tempting refuge for the naturalist. Not only did the pragmatism of Peirce, James and Dewey spawn Nagel's naturalism, but the differences between them seem thin: a difference in epistemology between prediction as sufficient or only necessary for knowledge. Adopting pragmatism will make short work of philosophical disputes – recall Nagel's condemnation of the realism/instrumentalism question quoted early in this paper. Moreover, pragmatism is so broad a church that it may even accommodate the social constructivist approach to science. It can accept that social factors help fix what counts as *successful* prediction and control, and these social factors may so vary as to hinder or aid its attainment. This is the lesson of Richard Rorty's version of pragmatism.

But by assimilating truth to warranted assertability pragmatism surrenders one of the central methodological responsibilities of science: the obligation to explain. In this case what apparently needs to be explained is, as Boyd pointed out at the beginning of the recent debate, the instrumental success of science. Pragmatism feels no obligation to explain science's success, because any such explanation will be entirely theoretical and will not enhance science's predictive success. Accordingly, by the lights of a pragmatic epistemology, explanations of the successes of science don't constitute knowledge. But real science does not share pragmatism's indifference to explanations which may add nothing immediately to our predictive power. And for this reason, naturalism cannot forgo such explanations either. Consider, if inference to the best explanation is a permissible form of reasoning in science – even when that explanation has no immediate predictive upshot – it must be as well warranted in naturalistic philosophy of science.

But now it appears we are in an infinite loop. Because this is, so to speak, where we came in, with the appeal to such inferences to underwrite realism. At this, the constructivist opponents of naturalism, whether empiricist or relativist, will take a page from Arthur Fine and insist that such principles of inference be independently justified – justified independently of naturalism.

Suppose the only thing the naturalist can say by way of reply to this challenge is that to ask a naturalist for the foundations of naturalism is question begging – naturalism is a foundational doctrine. Then we are at impasse. But if the naturalist replies by appeal to something more fundamental than science itself, the naturalist has given up its first axiom – no first philosophy. If we are not therefore to end the debate in a cul-de-sac of irreconcilable epistemic values, the naturalist will have to take philosophy seriously after all as prior or at least coequal guide to the nature of scientific knowledge. I am increasingly afraid that this is the only recourse for naturalists. Like others, their hope to solve or escape from the traditional questions of philosophy is doomed to be frustrated.

NOTES

Revised and reprinted by permission from *British Journal for Philosophy of Science*, 47 (1996): 1–26.

1. See Dennett, 1995, which argues that the theory of natural selection is a universal acid that can eat through all dogma.
2. Throughout this essay, the term anti-realism is used to name the doctrine that since our knowledge cannot transcend experience, we have no grounds to suppose that the theoretical claims of science are true. Anti-realism of the sort advocated by van Fraassen admits that theoretical claims are true or false, but advises agnosticism. Anti-realism in the philosophy of science must be distinguished from the doctrine of the same name associated with Michael Dummett, whose verificationism would in fact undermine views like van Fraassen's.
3. Boyd identifies a second feature of naturalistic epistemologies which he considers to be another "side of the same coin" as naturalism's rejection of *a priori* epistemology:

> The second essential feature of naturalism in epistemology is the...appeal to
> [unreduced] causal notions in the analysis of knowledge.... [T]he empiricist tradi-
> tion...has insisted on a non-realistic treatment of causal...powers: talk about causa-
> tion is thought to be reducible to talk about regularities in nature. Naturalistic episte-
> mology utterly breaks with this tradition: not only does it appeal to unreduced causal
> notions,...it employs such notions in the analysis of knowledge itself. [1980,
> p. 625]

Among the other naturalists whose views are canvassed here, none has committed themselves to any specific account of causation, still less one that turns its back on the attempt to analyze causal notions into more fundamental non-causal ones, like regularities, counterfactuals, possible worlds, universals, etc. It is also worth remembering that a regularity theory of causation is one easy to combine with the most robust realism, even Boyd's. The claim that "the abandonment of a priori stan-dards for the evaluation of inductive strategies and the employment of unreduced causal notions in the analysis of knowledge...are two sides of the same coin" [p. 625] is simply mistaken.

4. Van Fraassen allows for the literal (though unknowable) truth or falsity of theoret-ical statements, such as "electrons have quantized angular momentum." But such statements are no part of a canonical semantic model which constitutes *the* core of theory.

5. Besides, Laudan's pessimistic induction can identify cases where a theory made sig-nificant progress even though its degree of similarity and the respects in which it was similar to real systems was lower than its predecessor.

6. Darwin's is the most frequently mentioned name in *The Advancement of Science*. Laudan's is a close second.

7. Kitcher is not the only naturalist who needs this argument. In *Philosophical Natural-ism*, David Papineau struggles with the charge of circularity, and seems to deny its viciousness. But he also notes that "At some point in human history people acquired the ability, which they did not have before, to focus on certain kinds of explanations of physical phenomena, and to ignore others" [Papineau, 1994, p. 167]. An appeal to Darwinism here is just what the naturalist ordered to secure the foundations of inductive inference.

8. Boyd comes close to seeing the threat this argument poses, though he frames it as a threat only to realism, and not more broadly:

> Suppose...that the realist's explanation of the development of some field [of science]
> is advanced in defence of realism as a philosophical thesis. Plainly the resulting defence
> of realism is cogent only if the realist's explanation...is understood *realistically*. For
> example, only if the account of epistemic access to subatomic particles is understood
> realistically is the realist's case that atomic theory has an unobservable and theory-
> independent subject matter advanced. But on the realist's own account, her explanation
> and the account of epistemic access it incorporates are ordinary scientific theories them-
> selves grounded in the very research tradition regarding which a defence of realism is
> sought. To insist on a realistic interpretation of the realist's explanation would thus
> presuppose realism regarding the tradition in question. Thus the realist's appeal to her
> explanation of the development of instrumentally reliable methodology in an abductive
> argument for realism as a philosophical thesis is question-beggingly circular. [Boyd,
> 1989, p. 359]

Arthur Fine has pursued this argument against realism further than others. See Fine, in Leplin, 1984, and in Fine, 1986. But it is clear that what is at issue is the cornerstone of naturalism: the role of scientific findings to underwrite scientific methods used to reach those findings.

9. This is a gambit Boyd tries for his naturalistic realism:

> ... why would a realist philosophical package incorporating a realist version of that explanation be superior to [a constructive] empiricist package incorporating the explanation [for the instrumental reliability of science] instrumentally interpreted or as a constructivist package incorporating the realist's explanation understood as a piece of social construction? [Boyd, 1989, p. 388]

The answer to his question, says Boyd, is that the constructive empiricist package requires an infinite inductive regress, while the constructivist account begs the question by assuming "a distinctly constructivist conception of ... social construction" [p. 388]. In other words, Boyd defends naturalistic realism against the charge of circularity by pointing out the circularity or regress of arguments for its competitors.

2

Naturalistic Epistemology
for Eliminative Materialists

In this chapter I defend Quine's naturalistic epistemology [Quine, 1969], extend it and respond to its critics. In doing so I have borrowed freely from the work of philosophers who had no thought to defend or extend a Quinean naturalistic epistemology (hereafter QNE), and I have criticized the views of some exponents of (non-Quinean) naturalistic epistemology.

The main theses of QNE are well known: traditional epistemology, a largely or wholly a priori and foundationalist discipline, is to be replaced by an empirical inquiry which will be a "chapter" mainly of psychology. This compartment of psychology is to examine the relation between an epistemic agent's "meager sensory input" and its "torrential output" of descriptions of a three-dimensional world "in order to see how evidence relates to theory."

Naturalizing epistemology is an inevitable consequence of Quine's rejection of the analytic/synthetic distinction, his repudiation of a difference between a priori and a posteriori and his attack on modality. These three commitments, together and separately, lead through a variety of arguments to the conclusion that philosophy and science are continuous with one another: the former is just very general and abstract scientific theorizing. Accordingly, a compartment of philosophy that deals with psychological states, as epistemology traditionally does, should be viewed as very general and abstract psychological theorizing. Epistemology must be continuous with psychology because philosophy is continuous with science. If science cannot be a priori, neither can epistemology. If there are no truths in virtue of meanings, no appeal to meanings can underwrite epistemological claims, nor can there be a range of necessary truths to which we have epistemic access, and for which an epistemological account must be provided. Any validation of science will have to appeal to claims themselves contingent, because there are no other sorts of claims. To undercut QNE requires an onslaught on the central features of Quine's philosophy.[1]

1   QNE'S NORMATIVE DIMENSION

The most serious objection to QNE is that it isn't epistemology at all, because it surrenders the normative roles of epistemology. Epistemology is sometimes alleged to be normative on analogy with ethics. Both treat the grounds on which acts are justified or warranted. Ethics examines what we ought to do, epistemology what we ought to believe. Just as there is an is/ought distinction in ethics, there is one in epistemology. We cannot expect to extract normative epistemological conclusions from factual psychological ones. Accordingly, the objection runs, whatever QNE may be, it cannot count as epistemology.

But, in fact, epistemology is no more normative than, say, statistical methodology or engineering is.[2] Or so I shall argue.

Quine's error apparently is to have ignored the normative element which traditional epistemologists have recognized and which the naturalists among them attempted to explain non-normatively. Jaegwon Kim expresses this criticism forcefully:

> Quine . . . is asking us to set aside the entire framework of justification-centered epistemology. That is what is new in Quine's proposals. Quine is asking us to put in its place a purely descriptive, causal nomological science of human cognition. . . . [I]t is normativity that Quine is asking us to repudiate. Although Quine does not explicitly characterize traditional epistemology as "normative" or "prescriptive," his meaning is unmistakable. Epistemology is to be a "chapter of psychology," a law-based predictive-explanatory theory, like any other theory within empirical science; its principal job is to see how human cognizers develop theories ("their picture of the world") from observation ("the stimulation of their sensory receptors"). Epistemology is to go out of the business of justification. We earlier characterized traditional epistemology as essentially normative; we see why Quine wants us to reject it. Quine is urging us to replace a normative theory of cognition with a descriptive science. [Kim, 1993, p. 224]

In fact, in "Epistemology Naturalized" Quine argued that traditional epistemology should be replaced because it is a "first philosophy" – an a priori discipline, requiring analytical truths and justifying induction non-inductively. Quine's reasons for seeking the replacement of traditional epistemology do not include a desire to expunge its normative component. And this is as well, since epistemology does not have one. Epistemology is prescriptive in only a relatively unproblematical way. Its "normativity" is exhausted by the prudential force of the imperative of instrumental rationality. Thus it shares the strengths and weaknesses of this sort of rationality.

According to traditional epistemology the aim of inquiry is justified true belief. Epistemology is normative only to the extent that it seeks means to this end. QNE substitutes another end for inquiry to displace justified true belief, because it reflects doubts about the coherence of the concept of belief and the importance of truth to the aims of inquiry. QNE does not reject justification, however. It naturalizes it, like most other epistemologies.

According to QNE the goals of inquiry are, to a first approximation, prediction and control. Important qualifications on this goal need to be made, in the light of QNE's argument for why prediction and control are the goals of inquiry. QNE's argument for prediction and control as the goals of inquiry reflect its naturalism, in particular its commitment to Darwinian explanations.

Like most of our major functional components, our cognitive and perceptual economies have been and continue to be organized by blind variation and natural selection in order in the end to maximize fitness in an environment which apparently has remained unchanged in its most basic physical processes world out of time. Inquiry's immediate goal is instrumental for the mediate, long-term "intrinsic goal" of fitness-maximization. The fact that inquiry is organized to maximize fitness rules out epistemology's traditional goal, truth, pretty directly. Merely picking up truths is no assurance of survival.

QNE holds that there is at most only one intrinsic good, value or goal in nature: that is, there is only one goal which organisms strive for without its being a means to any other goal of theirs, but because it is a goal imposed on them by natural selection. Even this goal – the maximization of reproductive fitness – the goal to which all other goals in nature are instrumental – is an intrinsic good only in a rather technical sense. Fitness-maximization is a goal and does not subserve any other goals, because there are no other more fundamental goals in nature. The explanation of why biological systems aim at it does not involve any further goal. That biological creatures are fitness-maximizers is not explained functionally or teleologically. It is explained causally. If Darwin is right then in the end all our functional traits are shaped to attain this end, fitness-maximization, and it in turn has no further end. That is what makes fitness-maximization an intrinsic goal of organisms, in a purely naturalistic sense. Fitness-maximization's status as an intrinsic goal of organisms enables us to grade the means they employ to attain it for efficiency – i.e. instrumental rationality. It is in this grading that normativity enters: some means are more rational than others.

But the long-run teleology of fitness-maximization comes with enough short-term slack that many features of the human organism, including some (but surely not the most important) components of its cognitive economy may not be the direct result of selection for fitness-maximization, but may have emerged through drift, pleiotropy, or some other non-selective process operating within the constraints of a variegated environment. More to the point, our cognitive economy is not directly designed to maximize fitness. Indeed, hardly anything about us or other organisms for that matter is or has been directly selected for because of its immediate contribution to fitness-maximization. Like the rest of our adaptations, our cognitive economy is organized to attain much more specialized goals, which are instrumental for fitness-maximization. These more immediate goals, selected for because they are conducive to fitness-maximization, give epistemology its purely natural, prudential, instrumental normative force. Given our cognitive goals, establishing the best means to attain them may be complex but it will not be any more normative than the solution to a design problem in engineering or a linear programming problem in management theory. Once our epistemic goals are given, those epistemic acts are justified which secure these goals. Epistemic acts are justified just in case they are efficient means to epistemic goals. If these goals are themselves instrumentally rational for systems that aim at fitness-maximization, there is no non-natural normativity in epistemology.

It is crucial to bear in mind that naturalistic epistemology is not evolutionary epistemology! Little of immediate significance follows from the natural selection of cognitive systems either to psychologists or epistemologists. QNE recognizes that little follows for psychology and epistemology from the adaptational character of our cognitive and perceptual systems. The real work in naturalistic epistemology will have to be accomplished by empirical psychology. By itself the adapted character of the mind shows little about its structure and function because we do not know enough about the selective forces to infer back from the fact that we attain cognitive ends to the mechanisms by which we do so. In fact the only way to prove that our apparatus is optimally designed is to identify its immediate ends, its actual mechanisms, and its ambient environment in enough detail to show how the ends attained by these means within these constraints are in fact optimal. And this project requires the contribution of non-evolutionary, functional psychology. QNE is not evolutionary epistemology. In fact, it rejects at least some of its pretensions. But QNE does identify a biological function for thought strong enough to undercut the claim that epistemology is interestingly intrinsically normative.

What is the immediate goal of our cognitive economy? The exponent of QNE answers this question unequivocally enough. If the ultimate aim of cognition is fitness-maximization, then the more immediate goal is to attain those cognitive states which secure the means to fitness-maximization – survival to reproductive maturity, the ability to discriminate fittest available mating partners, etc. It is clear that a cognitive economy designed to enhance survival will do so only if it enables its owner to attain many other more immediate ends. The specific aim will be to attain those cognitive states most conducive to survival, *in something roughly like the order in which failing to act on them constitutes a threat to survival and acting on them enhances survival prospects*. Thus, nature will select for a cognitive economy to which "fire burns" is more salient than "plutonium is fissile."[3]

Many of the these salient cognitive states – beliefs – will be truths. But others will be falsehoods, justified by the fact that they are close enough to the truth to enhance survival, given the cognitive agent's environment. This is why the aim of our cognitive system cannot be acquiring the truth *tout court*. A multitude of truths are trivial, and a large number are useless until organisms in the species' line of descent develop or the environment changes in a way that gives them survival value. And many falsehoods will be good to believe just because they are close enough to truths valuable from the point of view of survival in a given environment. Nature will select for a cognitive economy that in the long run attains truths and successively closer approximations to the truth among the falsehoods it continues to believe. Moreover, as more and more features of the environment become salient for the cognitive agent's survival, evolution will favor those cognitive systems capable of detecting these features or approximations to them. Because a feature detected is a feature that can be exploited directly or indirectly to enhance survival. So it is a convenient approximation to identify the long-run goal of cognitive systems as the provision of (gradually less approximate) truths about the way the world is *in the order of their survival value* for such systems *modulo* their environments. A balder way of putting all this, familiar to real pragmatists, is the clam that the basic epistemic goal of our cognitive apparatus is prediction, and that beliefs are justified to the degree they meet this goal. But the point about the order of survival value is essential to identifying this goal.[4]

According to QNE, epistemology can get all the normativity it needs or ever needed, from evolutionary biology: those beliefs are justified which conduce to the intrinsic goal of fitness-maximizing via the instrumental goal of increasing predictive accuracy with respect to the order of truths salient for survival in the cognitive agent's environment. As between two beliefs, one is more well justified than the other, if, other things equal, its contribution to prediction with respect to salient truths is greater. Objection: this means that many falsehoods will be more well justified than many truths. True enough, but it is worth noticing that these more well justified falsehoods will approximate to truths highly important to survival, while the less well justified truths will be largely theoretical statements, less important for survival. If QNE were about knowledge in its traditional sense – "justified true belief" – this might be a serious defect. But QNE is not about knowledge, because it eschews beliefs. However, this is getting ahead of ourselves.

2 NATURALISM WITHOUT PLURALISM

It is of course obvious that our cognitive apparatus is employed for other ends than prediction, and this observation has led some naturalists to a more ecumenical goal for the apparatus itself than predictive success. In fact, many students of the nature of inquiry have embraced a doctrine they call naturalism just because it holds out the prospect of accommodating pluralism, if not relativism, about epistemic ends and goals. QNE needs to defend its monism if only to forestall relativism in the sciences. To do so, it must disallow pluralism. But first it must defend prediction and control (suitably qualified) as the sole end of inquiry.

QNE must respond to Philip Kitcher, for example, who argues that "Dissatisfaction with the purely pragmatic reduction of the problem [of identifying cognitive value] can be generated from thinking about scientific beliefs that make people feel uncomfortable . . . detract from their happiness, without offering any practical returns" [Kitcher, 1991, p. 103]. This is a traditional objection to pragmatism, advanced since the turn of the century by Lovejoy [1908], Russell [1910] and Rescher [1977]. Kitcher's version of the argument mistakes the single goal-type which shapes the kind of cognitive system agents have with the conflicting token-goals, epistemic and non-epistemic, that agents happen to have. The type-goal of happiness, like predictive success, is an instrumental goal important to survival. Sometimes tokens of these kinds of instrumental goals come into conflict. But this is hardly grounds to deny that prediction and control are the fundamental cognitive values. Compare: the function of the heart is to circulate the blood. This is a function, and not merely an effect of the heart's behavior, even when by pumping blood through an open wound, it leads to death. Similarly, prediction may be the cognitive system's goal even as it sometimes leads to the organism's unhappiness, and reduces its fitness. Teleological systems sometimes fail to attain their goals, both immediate and mediate, and sometimes they are employed with effects quite distinct from anything we would count as the goals which identify their function.

Kitcher's tentative conclusion is that "An adequate account of cognitive virtue should not simply be pragmatic. The aim of inquiry is not simply to enable us to anticipate the deliverances of experience (prediction) or to shape nature to our ends (control). The third goal of inquiry is scientific understanding" [p. 103].

In what does scientific understanding consist? Kitcher contrasts two accounts of the matter:

> One way is to adopt a strongly realist approach to explanation . . . the task of science is to expose the causal structure of the world, by delineating natural kinds and uncovering underlying structure.
>
> . . . An alternative approach avoids strong realism's commitment to independent natural kinds and causal mechanisms. Minimal realism can be developed by supposing that the assortment of objects and ordering of phenomena is driven by a search for unification. Understanding nature consists in achieving a unified vision of it . . . recognizing the role that categories [natural kinds] and causal dependencies [mechanisms] play in our ordering of the world. [pp. 104–105]

How would QNE argue against these alternatives? I say "would" because a full-blown treatment of "understanding" is beyond the scope of even a long paper. But the argument would proceed by claiming that the first alternative Kitcher offers us as a cognitive goal – uncovering causal structure – is unilluminating, and the second – explanatory unification – turns out to be an epistemic variant on the Naturalistic Fallacy in moral philosophy. Neither cognitive goal provides an alternative to epistemic pragmatism. Moreover, the pragmatist has a theory of understanding which will accommodate these immediate goals as means to the ultimate goal of predictive success.

According to Kitcher the realist's goal of inquiry is "to expose the causal structure of the world, by delineating natural kinds and uncovering underlying structure." But "expose," "delineate" and "uncover" may all turn out to be metaphors for "understand" and it is the nature of understanding which is in question. QNE holds that if understanding is anything at all, it is at least the having of predictively useful expectations about the environment. QNE treats understanding as exclusively a matter of successful prediction and control in part because of its Darwinism, and in part because of its eliminativism about the propositional attitudes that other accounts of understanding might appeal to.

But consider Kitcher's account of understanding. No exponent of "strong realism" he, Kitcher has long held that understanding is provided by *unification*. And Kitcher has provided some detailed examples of the items to be unified, and the relations among them that ever increasing unification can provide [Kitcher, 1984]. But, echoing Moore's question about any identification of the good with something else [Moore, 1903], the exponent of QNE will ask of unification, "Does unification provide understanding?" That is, what is it about the most unified set of beliefs that constitutes complete understanding? What is it about an increasingly economical compression of beliefs into more central and more derivative that makes it an improvement in understanding? Can it be mere definitional fiat or convention? No. There is something cognitively valuable in unification that everyone has recognized since Euclid. QNE provides a ready answer: unification is conducive to understanding, because it enhances the cognitive system's powers of prediction and control.[5] Unification is an important instrumental good. It is certainly not an intrinsic good, as Kitcher himself recognizes.[6] It will not even be an instrumental good in an environment which the strong realist would describe as extremely complex in its causal structure. For in such an environment increases in unification will come at the expense of exposing causal structure.

Of course given the most reasonable beliefs about the actual environment within which the cognitive system operates, unification is an invaluable near-term goal, one which is most well suited to achieving the ultimate predictive goals of the cognitive system. So, Kitcher's error of identifying increases in unification as the goal of inquiry is epistemically harmless, in the actual world.[7]

Focus on the actual world, and in particular on the local environment, is crucial. Like other forms of biology, QNE recognizes that adaptations are local, and that when the environment changes maladaptations can suddenly become valuable for reaching the cognitive system's goals. This makes epistemic optima environmentally relational. Traits advantageous in one environment will be disadvantageous in another, like skin color or sensitivity to noxious stimuli.

But relationalism is sometimes mistaken for relativism, and some epistemic pragmatists wrongly suppose that they are committed to relativism about cognitive value. Indeed, some embrace epistemic pragmatism because they think it will underwrite an attractive, ecumenical epistemic pluralism. Steven Stich illustrates this slide from the environmental relationalism of epistemic ends to some kind of epistemic relativism [Stich, 1990; Kornblith, 1994, quotations cited from the latter]. Stich describes his view as "normative cognitive pluralism," the thesis that there may be various systems of cognitive processes that are significantly different from one another, though they are all equally good, i.e. attain equally well what people take to be "intrinsically valuable" [Kornblith, p. 398]. Since values differ, the goals of inquiry can legitimately differ:

> If . . . people can and do intrinsically value a wide variety of things, with considerable variation from person to person and culture to culture, then pragmatic assessment of cognitive systems will be relative to highly variable facts about the users of those systems. . . . [G]iven the diversity of goals and values, it is all

but certain that different systems of cognitive processes will be pragmatically preferable for different people. [Kornblith, p. 401]

This sort of pluralism is based on a failure to distinguish the adaptational goals which have shaped all human and infra-human cognitive systems from the diverse sets of epistemic and non-epistemic goals of particular cognitive agents. QNE will insist that it is not what people take to be intrinsically valuable that shapes cognitive systems, or instrumentally justifies them. It is what nature makes instrumentally valuable for survival and fitness-maximization that does these things. This long-term goal of the cognitive system may and often does come into conflict with the short-term evaluations of individual cognitive agents in varying environments. But such conflicts do not underwrite any pluralism about epistemic aims.[8]

What QNE cannot allow is that "different systems of cognitive processes will be pragmatically preferable for different people" *in the same environment*. This monism about the aims of inquiry is one QNE requires to defeat real epistemic relativism. Real epistemic relativism is the thesis that there is no non-question-begging way to choose among methods of inquiry because there is no way to reconcile their alleged plurality of incompatible if not incommensurable goals. Cognitive pluralism must distinguish itself from epistemic relativism.

Different environments will select for differing means to attain the same long-term cognitive goals. QNE's latitudinarian attitude towards epistemic means is a direct consequence of its evolutionary character. The theory of natural selection recognizes that like other organs, cognitive systems are the result of quick and dirty solutions to adaptational problems, which start from what is available and make incremental improvements. It recognizes that not every feature of every system is always under selection. Some may not be selected for at all, except over the very long run. Natural selection appreciates that local environmental changes may move the evolution of cognitive systems through a trajectory that is far from a straight and narrow progress. QNE seeks to identify the specific forces that will explain the actual historical trajectory of cognitive systems. This part of QNE reflects findings and theories in evolutionary psychology (see for example the work of Tooby and Cosmides [Barkow, Tooby, and Cosmides, 1992]).

But latitudinarianism about means does not extend to relativism about ends. It is especially important for QNE to deal with the threat of real epistemic relativism (as opposed to Stich's kind). QNE declines to take the problem of Cartesian skepticism seriously as an obstacle to science or proto-science (common sense) for that matter. Quine famously argues that the illusions on which skepticism trades are themselves revealed by science, so that reassuring the skeptic is a job within science and not a prolegomena to it. But Quine early recognized the threat to epistemology from relativism in the academic study of science. For if science, or its philosophy, history, sociology and psychology, can show that our cognitive systems are not organized to attain some trans-historical goal, then naturalism and epistemic absolutism – the existence of a single goal of inquiry – are incompatible. In "Naturalistic Epistemology" Quine wrote:

> The dislodging of epistemology from its old status of first philosophy loosed a wave of epistemological nihilism. This mood is reflected in the tendency of Polanyi, Kuhn, and the late Russell Hanson, to belittle the role of evidence and to accentuate cultural relativism. Hanson ventured even to discredit the idea of observation, arguing that so-called observations vary from observer to observer with the amount of knowledge that the observers bring with them. . . . One man's observations is another man's closed book or flight of fancy. [Quine, 1969, cited from Kornblith, p. 29]

In "Epistemology Naturalized" Quine defended the integrity of observation to forestall relativism. But his prior development of underdetermination provides an apparently powerful argument for relativism even when relativism grants Quine recourse to an evidential base in observations. Moreover, historians of science, and philosophers of it like Laudan [1984], have argued that the actual goals of inquiry have shifted over the history of science in such a way as to cast doubt on QNE's candidate for the sole and ultimate criterion of epistemic justification.

QNE's commitment to prediction and control as overriding epistemic goals will be tolerant of the wrong turns and blind alleys in the relatively short history of fallible *Homo sapiens'* attempts to produce science. Like mutations, these wrong turns and blind alleys will reveal the components of our cognitive strategies and variations in it. As with other biological traits, the actual cognitive practices of scientists and *Homo sapiens* generally constitute a range of variations which will be random from the point of view of adaptation. Over the long run nature (and sometimes even human beings) will select for those modes of inquiry that successively increase fitness by enhancing prediction and control. The variation (i.e. relativity) of scientists' actually adduced criteria is the result of drift, and pleiotropy – the *selection of cognitive styles of* inquiry because they are hitched for the nonce to strategies that are *selected for*. (The distinction is Elliot Sober's [1984].) QNE's attitude to the history of science will be considered cavalier only by those who mistake the objectives of epistemology as a psychological (and ultimately biological) discipline with a geological time scale for the objectives of the history of science – an enterprise that begins with the Greeks. Naturalistic philosophers of science, Larry Laudan [Laudan, 1984] for example, have argued that science fails to show allegiance to any single methodology or goal for inquiry over time. But this is something QNE should lead us to expect. As biological and informational environments change, local selective forces will encourage some cognitive styles more than others, just so long as science as a whole continues to track enhancement in salient prediction and control.[9]

3 NATURALISM GOES ELIMINATIVIST

Kim writes:

> If justification drops out of epistemology, knowledge itself drops out of epistemology. For our concept of knowledge is inseparably tied to that of justification.... [K]nowledge is itself a normative notion. Quine's nonnormative, naturalized epistemology has no room for our concept of knowledge. It is not surprising that, in describing naturalized epistemology, Quine seldom talks about knowledge; instead he talks about "science" and "theories" and "representations." Quine would have us investigate how sensory stimulation "leads" to "theories" and "representations" of the world. I take it that within the traditional conception these "theories" and "representations" correspond to beliefs, or systems of beliefs; thus, what Quine would have us do is to investigate how sensory stimulation leads to the formation of beliefs about the world. [Kim, 1993, pp. 224–225]

Even if Quine were not an eliminativist about the propositional attitudes, this criticism would founder on the instrumental rationality that QNE employs to expound epistemic justification. But QNE is an eliminativist doctrine and has no qualms about an epistemology without knowledge.

In *Word and Object* [Quine, 1960], despairing of an extensional analysis of the propositional attitudes, Quine describes 'belief' and its cognates as part of an essentially "dramatic idiom." He famously compares his view with Brentano's: "One may accept Brentano's thesis either as showing the indispensability of intentional idioms and the importance of an autonomous science of intentions, or as showing the baselessness of intentional idioms, and the emptiness of a science of intentions" [Quine 1960, p. 221]. Epistemology however is not, in Quine's words, "drama"; it is science. Accordingly it will have no essential role for the contentful psychological attitudes.[10]

So, as Kim notes, it is not surprising that Quine seldom talks about "knowledge." But this is not for the reason Kim gives. The normativity of knowledge is not problematical for Quine. Rather it is the intentionality of knowledge that makes 'knowledge' an inconvenient notion in QNE. Knowledge is justified true belief, and belief is an intentional state.

Kim's argument for the normativity of knowledge proceeds from what he argues is the normativity of belief. Belief in turn is normative on Kim's view because the attribution of beliefs "ultimately requires 'radical interpretation' of the cognizer . . . assigning meaning to his utterances" on the Davidsonian [Davidson, 1984] assumption that the epistemic agent is a 'rational being', "a being whose cognitive 'output' is regulated and constrained by norms of rationality" [Kim, 1993, p. 228]. QNE can happily accept the holism that Davidson requires for beliefs. It is only the intentionality which prohibits QNE from describing justified cognitive states as knowledge. And even before developing this non-intentional epistemology, it is worth noting that nothing in Davidson's or other holistic accounts of the propositional attitudes requires anything more than instrumental rationality of believers.

But 'beliefs' are intentional. Therefore QNE must eschew the word, and the terms, like 'knowledge', which are defined in terms of it. Instead QNE can substitute terms like 'information' and 'understanding', and provide a non-intentional account of them in terms of prediction and control.

In fact, given QNE's biological bias, what it needs is as biological an introduction for the epistemic but non-intentional terminology of Naturalized Epistemology as possible. If Jerry Fodor is correct, this is a need which some philosophers of psychology have inadvertently been filling for some time. These philosophers, among them Dennett (since Dennett, 1969), Bennett (unnoticed by Fodor but since Bennett, 1976), Millikan (since Millikan, 1984), Israel (since Israel, 1987), but not Dretske (see Dretske, 1979, pp. 205ff.), have attempted to offer accounts of intentional content by appeal to natural selection of among a large but blind variation of neural states on the basis of their appropriateness to efferent environmental contingencies: appropriateness is a matter of afferent behavior that is conducive to meeting the organism's goals – long term and short. Fodor rejects these accounts of intentionality because they fail to capture the intensionality of mental states. These accounts show at most how neural states can carry information, but not how they can have content. (Fodor, 1990, Part One. Fodor apparently independently rediscovers an argument advanced in Rosenberg, 1987, Part One, and Rosenberg, 1989.) But one philosopher's *modus tollens* is another philosopher's *modus ponens*. QNE requires an account of cognitive states which makes them susceptible to instrumental justification independent of intentional content, if

any. An account of how they may have informational content without intensionality (and thus without intentionality) is just what QNE requires.

Fodor rightly points out that

> appeals to mechanisms of selection won't decide between *reliably equivalent*
> content ascriptions; i.e. they won't decide between any pair of equivalent con-
> tent ascriptions where the equivalence is counterfactual supporting. To put this
> in the formal mode, the context: *was selected for representing things as F* is
> transparent to substitutions of predicates reliably coextensive with F. A fortiori,
> it is transparent to the substitution of predicates *necessarily* (including *nomo-
> logically* necessarily) coextensive with F. In consequence, evolutionary theory
> offers us no contexts that are as intensional as 'believes that . . .' If this is right,
> then it's a conclusive reason to doubt that appeals to evolutionary teleology can
> reconstruct the intentionality of mental states. [Fodor, 1991, Part One, p. 73]

What a relief! Intentionality is the last thing QNE wants for our cognitive
processes. But what it needs for purposes of epistemology is the identification
of those cognitive states which may be more or less well justified, and the
conditions under which one such state will be more or less well justified than
another.[11]

In *Linguistic Behavior* [1976] Jonathan Bennett attempts naturalistically to
upgrade non-intentional cognitive states into beliefs. His analysis falls afoul
of the very criticism Fodor advances just because it exploits the selectionist
program. Below I sketch his naturalistic account of intentional content and
show how the services it can perform for QNE are not limited by its failure
to provide an account of belief. The same silver lining is to be found in other
failed attempts to cash intentionality in for environmental sensitivity. I employ
Bennett's in part because his theory is developed with an eye to psychology,
not epistemology, and so helps QNE cash in its claim that epistemology comes
out of a branch of psychology.

Consider organisms whose behavior is thoroughgoingly appropriate to
their local and global environments. Call them animals. QNE has no qualms
about the existence of such teleological systems, because it can help itself to
several wholly extensional accounts of goal-directedness from Nagel [1977]
to Wright [1976]. Like the rest of biology QNE recognizes that identifying the
goals, ends, and purposes which help explain behavior is a highly fallible matter of empirical enquiry. Nevertheless, we can identify many of an organism's
goals and those behaviors of the organism causally necessary and sufficient
in the circumstances for attaining its goals. It is crucial to Bennett's program
and to QNE that our description of these goals be transparent to the substitution of predicates and terms that are, as Fodor says, "reliably" coextensive

or coreferential. But this is not a problem: As Fodor says so trenchantly, *"Darwin cares how many flies [a frog] eats, but not what description [it] eat[s] them under"* [Fodor, 1991, p. 73, emphasis in original]. By "reliable" Fodor means "counterfactual supporting"; QNE can accept this meaning so long as it provides a non-modal account of counterfactuals, or it can provide another non-epistemic, non-intentional account of "reliable" coextensiveness or co-reference. (See note 11 below.)

With extensionally identified goals in hand, Bennett introduces the non-intentional building block for beliefs. QNE will encourage him to build it up to capture as much of the notion of belief as possible without capturing its intensionality. Since Bennett fails to capture intensionality, QNE can help itself fully to the fruits of Bennett's entire upgrading project. Bennett's key building block is the concept of 'registration': 'x registers that P' is a "theoretical term, standing for whatever it is about a given animal which validates predictions of its behavior from facts about its environment" [Bennett, 1976, p. 52]. When does x register that P?

> If *a* is in an environment which is *relevantly similar* to some environment where P is *conspicuously* the case, then *a* registers that P. The italicized expressions are shorthand jargon: each must be defined through [a] . . . combined goal and registration theory for *a*. That theory contains hypotheses about which features of its environments *a* is sensitive to, and those hypotheses pick out a certain class of environments which is made up as follows: (1) a core of environments where P is the case, and where the features because of which P is the case are ones to which *a* is sensitive [the cases of conspicuous P] and (2) a surrounding ring of environments where P is not the case but which do not differ, in any respects to which a is sensitive, from some members of the core [these are the relevantly similar environments]. [p. 54]

Organism *a* is sensitive to an environmental feature just in case *a*'s teleological behavior varies frequently enough with the presence of the feature so that when present the feature is a useful one for predicting *a*'s behavior.

Of course being sensitive to an environmental feature involves registering its presence. No one said the introduction of the theoretical term "registration" was going to be a matter of reductive definition. The only stricture on the term's introduction is that it be non-intentional.

Bennett upgrades registration into belief by introducing the notion of "educability" and then claiming that animals that register *P* and are highly educable with respect to *P* (and many kinds of propositions of the same kinds as P) also believe that *P*. Educability is defined in terms of two behaviorally specifiable properties of animals: Call animal *a* educable with respect to *P* if

47

1. a registers that P in environments $K_1, \ldots, K_n$, that is, a evinces behavior appropriate to P's being the case in these environments. Suppose, however, that P is not the case in K_n. Then eventually a ceases to evince P-appropriate behavior in K_n. The less time this takes, the more educable a is.

2. a registers that P in environments $K_1, \ldots, K_n$, but not in K_{n+1}, although P is the case in K_{n+1}. Then eventually a begins to evince behavior in K_{n+1} appropriate to P's being the case in it. The less time this takes, the more educable a is. [Adapted from Bennett, 1976, p. 84]

It is a fatal defect of Bennett's analysis of belief that educable registrations are not sensitive to the substitution of co-extensive predicates or co-referring terms: If a registers that b is F, and $b = c$, then a registers that c is F, and if $(x)Fx \rightarrow Gx$ is a law, then a also registers that b is G. But it is a virtue so far as QNE is concerned. For it will encourage an epistemological study of highly educable registrations that substitutes for traditional intentional epistemology, and makes substantial use of empirical psychology.

4   THE EPISTEMOLOGY AND PSYCHOLOGY OF REGISTRATIONS

What will QNE's program for an epistemology of registrations look like? Of course registrations are not essential, just illustrative. If empirical psychology reveals that there is a more suitable notion than "registration" for systematizing the cognitive processing of animals, QNE is free to adapt it.

One primary task for psychology is to catalogue the types of registrations of which animals of varying species are capable, whether they are modularized in Fodor's [1984] sense and Quine's [1969], how they co-vary, how long registrations can exist and continue to control output, and the degree to which their registrations are educable. This is a task already long under way in behavioral biology, and areas of experimental psychology like psychophysics, and behavioral learning theory, psychological concerns that were going, and going well, when Quine first broached QNE.

But Quine's sixties' behaviorism is no essential part of QNE. Empirical psychology has moved from behaviorism to cognitive science, neuroscience – computational and otherwise, computer simulation, artificial intelligence, etc. Any and all of these compartments of psychology can and do tell us much about what and how organisms do and can register propositions about their environments. These subdisciplines transcend the behaviorism Quine endorsed by shedding light on the mechanisms animals employ to acquire and educate registration. QNE is no more lumbered by behaviorism than psychology itself is, and its only stricture on theorizing about registrations is that it ultimately or immediately frees itself from the burdens of intentionality. QNE is that compartment of non-intentional empirical psychology which exploits its findings to frame hypothetical imperatives about how cognitive systems can most effectively attain their instrumental and intrinsic aims.

As yet many of these psychological studies are tainted with intentional attributions. But even when this happens, these attributions are not taken very seriously outside of the human case at any rate. Humans vocalize propositions, and we take these vocalizations to identify unique propositions they believe. Since animals lack language, there is no matter of fact about exactly which among nomologically or otherwise reliably equivalent sets of propositions they believe. QNE insists that intentional attributions be taken no more seriously in the human case as a guide to what exactly is believed. For the degree of indeterminacy in attribution remains very great in the human case. It is only the human use of language that gives a false sense that there is exactly one proposition to which the psychological attitude is directed. This is the relativistic point of radical translation, and any version of QNE will take it seriously.

By contrast, though there will be in many cases great indeterminacy about what an animal, human or infra-human registers, in the long run there will be a fact of the matter about it: an animal registers just those propositions required to explain (all of) its goal-directed behavior.

Humans have extremely educable registrations and large numbers of them. Doubtless the linguistic behavior to which they lead is largely responsible for suggesting the intentional vocabulary of folk psychology and traditional epistemology to common sense. The observable evidence on which QNE and folk psychology-*cum*-traditional epistemology make their respective attributions of registration and belief is exactly the same. But where folk psychology misleadingly appears to warrant the attribution of one specific proposition as believed ("Water freezes at 0°"), QNE warrants a large class of registrations of nomically and otherwise reliably correlated propositions ("Water freezes at 0° Celsius," "H_2O expanded below 32° Fahrenheit," "The substance on which pH-meters are calibrated becomes a solid at 273° Kelvin," etc.) so long as each of the propositions registered makes a contribution to the explanation of behavior. The relatively larger number of propositions registered by an organism than it is commonly supposed to believe reflects QNE's avoidance of the ambiguity of intentional attribution. No amount of behavior will indicate exactly what an intentional agent believes (whence indeterminacy of translation). But finite amounts of behavior will indicate at least some of the propositions an animal registers.

The business of psychology is to identify which sets of registrations varying animals can and do register under differing environments, given changing
needs and goals. It is worth noting that registrations provide a sort of "narrow
content" that honors Fodor's [1981] principle of methodological solipsism
and Stich's [1983] principle of autonomy. This is important because, as both
Stich and Fodor [before Fodor, 1994] have argued, broad intentional content
of the sort that, say, Burge [1979] advocates, makes both psychology and QNE
impossible. Psychology thus provides an account of narrow contentful registrations in various species-in-environments (including laboratory settings).
The business of epistemology is to identify which of them is (instrumentally)
justified for the animal in its environment.

Since registrations are propositional, a certain amount of traditional epistemology can be taken over by QNE without significant alteration. For example,
all the inferential rules of deductive logic hold for registrations: if a registers
that P and a registers that *If Q then Q*, then a will be justified in registering
that Q. But just because standard principles of logic hold for registrations,
it does not follow that QNE enjoins them as instrumentally rational in all
contexts. As Cherniak [1986] points out, some relatively obvious hypothetical imperatives, like "avoid inconsistencies," will be impossible for cognitive
systems capable of registering many propositions to act on. QNE will help
explain (in some suitably non-intentional way) and (instrumentally) justify if
and when such an imperative makes sense.

By contrast, the instrumental injunctions of the Bayesians not to make
Dutch books, and to update the strength of a probabilistic registration by
employing Bayes' theorem, is readily accommodated by QNE. Once registrations are introduced, there is no difficulty constructing the notion of x
registers that the probability of P is n. All we need is a naturalistic non-
intentional substitute for a's preference ranking. A ranking of animal a's
goals – long-, medium-, and short-term, at time t, is something we can expect a non-intentional biology to give us. Suppose a's prior behavior or its
species-specific dispositions reveal that a ranks goal G' above goal G''. And
suppose it registers that behavior B' will enable it to attain G' with unknown
probability n' and B'' will enable it to attain G'' with unknown probability n''.
If it engages in behavior B' aimed at lower ranked G' and not B'' aimed at
higher ranked G'', then the difference between a's subjective probabilities n''
and n' must be greater than the difference between its rankings G'' and G'.
[For further discussion, see Bennett, 1976, pp. 69–70.]

QNE's introduction of subjective probabilities can start well down on the
phylogenetic ladder. The lower down it starts, the coarser the probability
assignments it will be able to make, or need to make for that matter. The

more complex the behavior, the more complex the registrations it will need to attribute and the harder it will be to infer the goal structure which together with behavior enables us to attribute subjectively probabilistic registrations. But these are problems that non-intentional psychology and the rest of science are not constitutionally debarred from helping QNE to solve.

Up the phylogenetic ladder, and over the history of the cognitive development of our species, adaptational needs combined with prior registrations will structure the priority of goals. And it will multiply them as human life becomes more complicated so that probabilistic registrations can be assigned with increasing precision. Once natural science has developed far enough so that predictive success in the laboratory and the field becomes an adaptation, and scientists register the predictive successes of their predecessors, QNE will have no trouble accounting for the advancement of science, naturalistically.

QNE's problems in applying rational decision theory will never be as intractable as those which the indeterminacy of preference/expectation attributions raises for non-naturalistic Bayesian epistemology. The intentional circle of "radical interpretation" prohibits the theory of decision under uncertainty from distinguishing a wide range of desire-belief packages from one another because of their equivalence of effects in behavior. And traditional epistemology's intentionality debars non-intentional science from helping to reduce this indeterminism. QNE suffers from neither of these problems.

The insulation of intentional epistemology from empirical psychology is illustrated by its reaction to studies which purport to show that judgment under conditions of uncertainty systematically violates the strictures drawn from the probability calculus [see Kahneman, Slovic, and Tversky, 1982]. Where traditional epistemology has difficulty reconciling itself to heuristic judgment strategies, QNE has no difficulty exploiting these findings to enrich epistemology. Identifying the inferential shortcuts (from registrations to registrations, of course, not from beliefs to beliefs) that, though generally invalid, work effectively for *Homo sapiens* in specified *environments*, is a paradigmatical enterprise of QNE-*cum*-psychology. Rationalizing their incompatibility with "orthodox" principles of inference [Cohen, 1981] is a simple matter for QNE. Appropriate strategies of inference will vary over cognitive systems and environments: differing systems in the same environment may use differing inference-rules with equal effectiveness. Similar systems in differing environments will probably do so.

Nor need QNE embrace the view, of those who deny human rationality like Stich [1984], that experiments which reveal these heuristic reasoning patterns "demonstrat[e] . . . that human reasoning often deviates substantially

51

from the standard provided by normative canons of inference" [Stich, 1984, quoted from Kornblith, 1994, p. 342]. For QNE normative canons of reasoning are the ones which conduce to the attainment of epistemic ends. These will vary over environments, and in some environments, heuristics that violate traditional canons will turn out to be normatively well-justified.

Similar considerations blunt Stich's criticism of some arguments that might reconcile the persistent employment of invalid heuristics with the attribution of rationality. Stich attributes to Dennett [1981] the view that "no experiment could demonstrate that people systematically involve invalid or irrational inferential strategies" [p. 343]. In brief, the ground for this conclusion attributed to Dennett is that intentionality requires (instrumental) rationality, and humans are intentional systems. But once (instrumental) rationality's adaptationalist character is properly understood, there is no reason to suppose that traditional epistemology's "normative canons of inference" will be justified in all environments for all cognitive systems. Accordingly, neither QNE nor the (instrumental) rationality approach to intentional attribution need resist the conclusion "that people systematically invoke invalid or irrational inferential strategies."[12]

This combination of valuing a cognitive process for its survival value while admitting its failure to accord with canons of traditional epistemology is not new to QNE. Already in "Naturalistic Epistemology" [Quine, 1969, quoted in Kornblith, 1994, p. 30] Quine was keenly aware of the role of color perception in survival, while in "Natural Kinds" [Quine, 1969, quoted in Kornblith, 1994, p. 66] he denied colors the status of natural kinds. Empirical science and philosophical awareness of it have vindicated Quine's position as prescient. Beginning with the scientific work of Edwin Land [1977], Hardin's study of color [Hardin, 1988] both exemplifies the continuity of psychology and epistemology and underwrites Quine's earlier denial that colors are natural kinds. Hardin shows how developments well after 1969 (the date of "Epistemology Naturalized") have suggested that perceived "colors have no counterparts in the physical world," and yet it is "advantageous for animals like us to represent the world to ourselves as colored" [Hardin 1992]. Hardin points out that colors are neither wavelengths, nor spectral reflectances, or any other of the physical features of the environment that cause color perceptions in us. This is because for humans there are just four basic unitary perceptual hues. But there is no set of exactly four particular basic spectral frequencies of which other wavelengths are composed. "The perceptual structure of colors thus has no counterpart in the domain of wavelengths of light even though we see those colors because we are stimulated by light that has an appropriate wavelength configuration" [Hardin, 1992, p. 371].

But, responding to the world as if it were composed of colored objects gives us an advantage in coping with our environment: "By representing the reflectances in the environment by means of basic visual qualities analogous to what we humans see as red, yellow, green, or blue, or pairwise combinations of these, animals like ourselves that have restricted information processing capabilities are better able to go about our business of recognizing objects and receiving signals from other organisms" [Hardin, 1992, p. 380].

To distinguish something among objects of similar lightness, our receptive system must be sensitive to small wavelength differences. To receive and discriminate signals, our sensory system needs to employ a small number of basic categories that ignore minor variations among stimuli, classifying them into a small number of salient categories. Hardin concludes:

> The system of qualitative classification that this involves need not match any analogous set of structures outside the organism in order to provide real advantages to the animal that uses it. The hues that we human beings see express our system of coding wavelength information rather than some set of properties of reflecting surfaces. . . . [It] supplies the means by which a rich amount of sensory information can be rapidly and efficiently represented by cognitive machinery of limited capacity. [p. 381]

This is QNE at its best, ironically enough revealing the relation between "torrential *input*" and "meager *output*," and showing how our theory of a colored world transcends available evidence but is justified in doing so.

5   BUT IS IT EPISTEMOLOGY?

We are now in a position to assess a last objection to QNE advanced by Kim. QNE's response to this objection summarizes a good deal of the exposition and defense laid out in this paper. Kim briefly considers and dismisses the eliminativism on which much of QNE rests. First Kim sketches an eliminativist response to his objections to QNE:

> It might be thought we could simply avoid [the-normativity-of-belief-problem] stemming from belief-attributions by refusing to think of cognitive output as consisting of "beliefs," namely as states having propositional contents. The "representations" Quine speaks of should be taken as appropriate neural states, and this means that all we need is to be able to discern neural states of organisms. This requires only neurophysiology and the like, not the normative theory of rational belief.

Kim's reply proceeds as follows:

> either the "appropriate" neural states are identified by seeing how they correlate with beliefs, in which case we still need to contend with the problem of radical interpretation, or beliefs are entirely bypassed. In the latter case, belief, along with justification, drops out of Quinean epistemology, and it is unclear in what sense we are left with an inquiry that has anything to do with knowledge. [Kim, 1993, p. 230]

But now we see Kim's view of the alternatives to intentional psychology is too narrow. The first alternative to an intentional psychology is to try to develop another functional or teleological theory which is free from intentions. Any one of the near-miss attempts to naturalize the intentions may provide the basis for such a psychology. I have sketched one. If it succeeds QNE need not identify the appropriate neural states in order to undertake its mission of a scientific epistemology.

Radical interpretation is not intrinsically normative in any case. Even if radical interpretation were intrinsically instead of just instrumentally normative, we would still have a notion of justification. But it would certify understanding as a non-intentional achievement, and there would still be a good deal to learn about which of our registrations are justified and when. This is surely an enterprise recognizable as epistemology, even when naturalized.

NOTES

Reprinted with permission from *Philosophy and Phenomenological Research*, 59 (June 1999).

1. Quine does not argue for QNE from these premises. He proceeds from the failure of the empiricist's program to "validate science" as knowledge founded on certainty about foundations. This rightly opened Quine to the objection that the foundationalist program does not exhaust traditional epistemology and the failure of phenomenalist reduction of all material object claims to ones about sensations does not herald the demise of non-natural epistemology. For examples of this criticism see Sober, 1978; Kim, 1993; Bonjour, 1995.
2. Quine makes the comparisons briefly in 1990, pp. 20, 21. The analogy between epistemology and ethics is unconvincing, both as a matter of history and of philosophy. To begin with, by contrast with ethics, there is a longer and more majoritarian tradition of attempts to cash in the normative content of epistemic claims for non-normative, purely descriptive ones. It begins with Descartes's attempt to analyze justification in terms of first-person sensory incorrigibility. Indeed about the only exceptions to this tradition in recent times appear to be Roderick Chisholm [1977, 2d. ed.] and philosophers he has influenced. If we are pursuing the analogy with ethics, the majority tradition in epistemology is in fact properly to be construed

as a naturalistic one. In the history of ethics, by contrast, naturalism is much more of a minority view.

3. For the moment let us refer to the relevant states of these cognitive systems as "beliefs," and help ourselves to other terms from traditional epistemology and intentional psychology, knowing full well that QNE will have to find substitutes for them, if it is fully to reflect Quine's eliminativist program. See section 3 below.

4. Quine himself wrongly waffles on this point in a way that weakens the case for normative epistemology seriously:

> When I cite prediction as the checkpoints of science, I do not see that as normative. I see it as defining a particular language game, in Wittgenstein's phrase: a game of science, in contrast with other good language games such as fiction and poetry . . .
> [P]rediction is not the main purpose of the science game. It is what decides the game, like runs and outs in baseball. It is occasionally the purpose, and in primitive times it gave primitive science its survival value. But nowadays the overwhelming purposes of the science game are technology and understanding. [1990, p. 20]

Technology, however, is but prediction applied, and sits uneasily with understanding as an epistemological goal, as I argue below.

5. This answer will have to be revised when QNE comes to substitute for the unwanted intentional term 'knowledge'.

6. See Kitcher, 1993, p. 241, quoted in the next footnote. Kitcher argues that our faculty of understanding is the result of natural selection for one which seeks significant truths.

7. Ironically, Kitcher himself is committed to a very strong naturalism, one which not only minimizes the epistemic goal of truth and approximation to it in favor of what he calls "significance," and he is committed to a very strongly Darwinian thesis about how evolution has shaped cognitive capacities for survival via the acquisition of significant beliefs. *In The Advancement of Science* he writes:

> As a consequence of our genotype and our early developmental environments, human beings come initially to categorize the world in a particular way, to view certain kinds of things as dependent on others, to generalize from single instances of especially salient types. Moreover, . . . there is a propensity to restrict those generalizations in particular ways when matters go awry. . . . [T]his *primitive apparatus* works tolerably well in confronting the problems our hominid ancestors encountered; it is relatively well designed for enabling primates with certain capacities and limitations to cope with savannah environment and with the complexities of a primate society. [Kitcher, 1993, p. 241]

The last clause seems a barely acknowledged admission that our cognitive capacities have been selected for "coping," that is, for prediction and control. Unification is a way of coping in an environment like the African savannah and its successors.

8. According to Stich, one way in which "normative cognitive pluralism might arise" does turn on the relational character of *local* evolutionary adaptation – the relativization of immediate epistemic goals to the environment. His mistake is to conclude that this relationalism makes for epistemic pluralism or relativism:

> Suppose that our account of what makes one system of cognitive processes better than another involves relational features of the systems being evaluated, where one of the elements in the relation is the person or group using the system, or some property of that

person or group. If systems of cognitive processes are evaluated in this user-relative way, then in general it will not make sense to ask whether one system is better than another (full stop). Rather we must ask whether one system is better than another for a given person or group, while another system is better for another person or group. I take this possibility to be the hallmark of relativism in the assessment of cognitive processes. An account of cognitive evaluation is relativistic if the assessments of cognitive systems it offers are sensitive to facts about the persons or groups using the system.

... The likelihood of a given system leading to certain consequences will generally depend on the environment in which the system is operating. If we keep the goals constant, the probability that a given system of cognitive processes will lead to these goals is going to depend, to a significant extent, on the circumstances of the person using the system. [Kornblith, pp. 400–401]

QNE keeps these goals constant, or rather reads the constant long-term goal off from the adaptational forces which are slowly but surely shaping human and infra-human cognitive systems. But it recognizes that the short-term goals, and indeed the medium-term goals and the means of meeting them, will vary over environments in the way that all other adaptations do.

9. Remarkably enough QNE can appeal to Larry Laudan for concurrence in QNE, though Laudan has Quine slightly wrong:

> Whereas Quine evidently sees naturalized epistemology as constituting a subfield within epistemology, I [Laudan] argue that an empirical epistemology or methodology is neither a part of, nor subordinate to, psychology. And typically , it would draw more heavily on physics and biology than on psychology. . . .
>
> Once we realize . . . that methodological norms and rules assert empirically testable relations between ends and means, it should become clear that epistemic norms, construed of course as conditional imperatives (conditional relative to a given set of aims), should form the core of a naturalistic epistemology. [Laudan, 1984, p. 41]

Laudan's exigetical error is failing to acknowledge that for Quine psychology is a thoroughly biological science. All that is required of him to share in QNE is the admission that the various historically actual aims of inquiry are themselves justified or not by the degree they are or are not instrumental in relation to their times (their environments) for attaining the biological aim of survival. This Whiggish interpretation is important for QNE's refutation of relativism in the philosophy of science. It will underwrite QNE's historical explanation of scientific change as progressive (if not linear) improvement in predictive precision and range.

10. There is some backsliding on this in Quine, 1990, pp. 70–72, in which anomalous monism is endorsed – nomological danglers and all. But once endorsed, Quine denies the intensional anything but the potential heuristic, instrumental role of setting us scientific problems to be solved extensionally. But what is sauce for extensional science, is sauce for extensional epistemology – a compartment of science.

11. Some will object to QNE's eliminativism on the ground that it is already committed to intentionality by its commitment to laws and theories, like Darwin's and others', it will require to make epistemology part of science. Dretske [1981] argues that nomological connections are intentional. This, he writes, "suggests that we may be able to understand the peculiar intentional structure of our cognitive states as symptoms of their underlying [nomological] information-theoretic character" [p. 76]. If this is correct, QNE cannot forgo intentionality unless it is prepared to forgo appeal to nomological force. Thus, it cannot adapt Fodor's argument as a

modus tollens to the conclusion that epistemology can proceed without intentionality. But exponents of QNE should reject this line of reasoning for several reasons. First, according to Dretske, nomological relations are intentional because they are intensional, and intensional because they are modal [p. 77]. But modality is different from intentionality even if intensionality is symptomatic of both. Second, the intensionality of nomological and causal contexts is independently questionable [see Rosenberg and Martin, 1979]. Third, powerful analyses of modality, like David Lewis's [1976, 1989], are free from intentionality. Indeed, Lewis's account of modality is designed to satisfy Quinean strictures. However, all QNE really requires is that the intensional or non-truth-functional character (if any) of general laws have its sources elsewhere than in the intentionality that characterizes psychological states.

12. It should be clear that what stands between QNE and Dennett is not the role of (instrumental) rationality in the explanation of behavior, but whether its role is sufficient for intentionality, as well as necessary for it. See Dennett, 1991.

3

Limits to Biological Knowledge

In this chapter I want to argue that one particular science faces limits that do not confront other sciences, and that these limits reflect a combination of facts about the world and facts about the cognitive and computational limitations of the scientists whose business it is to advance the frontiers of this science. The science is biology, and the limitations I claim it faces are those of explanatory and predictive power. In the first part of this chapter I advance a contingent, factual argument about the process of natural selection which destines the biology in which we humans can take an interest to a kind of explanatory and predictive weakness absent in our physical science. I then go on to show how these limitations are reflected in at least two of the ruling orthodoxies in the philosophy of biology: the commitment to the semantic approach to theories, and to physicalist antireductionism.

If I am correct about the limits to biological knowledge, we must face some serious issues in our conception of what scientific adequacy and explanatory understanding consist. My claim is that biology is far more limited in its ultimate degree of attainment of scientific adequacy than are the physical sciences, because the only generalizations of which biology is capable will not provide for the sort of coordinated improvement in explanation and prediction which characterizes increasingly adequate science. This fact about biology reflects as much on the biologist as it does on the phenomena the biologist seeks to explain and predict. Were we much smarter, physics and chemistry would remain very much as they are, but biology would look much different. The history of chemistry and physics is a history of an increasingly asymptotic approach to the closed cycle of theory-driven improvement in prediction which make us confident that we are uncovering the fundamental regularities that govern the universe. If we were smarter – if we could calculate faster and hold more facts in short-term memory – quantum mechanics, electromagnetism, thermodynamics, physical and organic chemistry would still be characterized by physical theories recognizably similar to the ones we know and love.

But if we were smarter, then the generalizations of biology as we know it would give way either to others, couched in kind-terms different from those which now characterize biology, or they would give way to the laws and theories of physical science aided only by a principle of natural selection. Were this to happen, the limitations on attaining the closed cycle of theory-driven predictive improvements about the phenomena we now call biological could be abridged. But the resulting science would not be recognizably biology. And besides, we aren't going to transcend our natural cognitive and computational limits any time soon.

The reason for the predictive weakness and consequent explanatory limits of biology is of course to be found in the generalizations of which it is capable. To see why biology is capable only of weak generalizations we need only reflect on two considerations: the mechanism of natural selection and the inevitability for us of individuating biological kinds by their causal roles.

Natural selection "chooses" variants by *some of their effects*, those which fortuitously enhance survival and reproduction. When natural selection encourages variants to become packaged together into larger units, the adaptations become functions. Selection for adaptation and function kicks in at a relatively low level in the organization of matter. As soon as molecules develop the disposition chemically, thermodynamically or catalytically to encourage the production of more tokens of their own kind, selection gets enough of a toehold to take hold. Among such duplicating molecules, at apparently every level above the polynucleotide, there are frequently to be found multiple *physically distinct* structures with some (nearly) identical rates of duplication, different combinations of different types of atoms and molecules, that are close enough to being equally likely to foster the appearance of more tokens of the types they instantiate. Thus, so far as adaptation is concerned, from the lowest level of organization onward there are frequently *ties* between structurally different molecules for first place in the race to be selected. And, as with many contests, in case of ties, duplicate prizes are awarded. For the prizes are increased representation of the selected types in the next "reproductive generation." And this will be true up the chain of chemical being all the way to the organelle, cell, organ, organism, kin-group, etc.

It is the nature of any mechanism that selects for effects, that *it cannot discriminate between differing structures with identical effects*. And functional equivalence combined with structural difference will always increase as physical combinations become larger and more physically differentiated from one another. Moreover, perfect functional *equivalence* isn't necessary.

Mere functional similarity will do. So long as two or more physically different structures have packages of effects each of which has roughly the same
repercussions for duplication in the same environment, selection will not be
able to discriminate between them unless rates of duplication are low and
the environment remains very constant for long periods of time. Note that
natural selection makes functional equivalence-*cum*-structural diversity *the
rule and not the exception* at every level of organization above the molecular.
In purely physical, or chemical processes, where there is no opportunity for
nature to select by effects, structural differences with equivalent effects are
the exception, if they obtain at all.

Now, if selection for function is blind to differences in structure, then there
will be nothing even close to a strict law in any science which individuates
kinds by selected effects, that is by functions. This will include biology, and
all the special sciences that humans have elaborated. From molecular biology
through neuroscience, psychology, sociology, economics, etc., individuation
is functional. That cognitive agents of our perceptual powers individuate
functionally should be no surprise. Cognitive agents seek laws relating natural
kinds. Observations by those with perceptual apparatus like ours reveals few
immediately obvious regularities. If explanations require generalizations, we
have to theorize. We need labels for the objects of our theorizing even when
they cannot be detected, because they are too big or too small, or mental.
We cannot individuate electrons, genes, ids, expectations about inflation, or
social classes structurally because we cannot detect their physical features.
But we can identify their presumptive effects. This makes most theoretical
vocabulary "causal role"-descriptions.

It is easy to show that there will be no strict exceptionless generalizations
incorporating functional kinds. Suppose we seek a generalization about all
F's, where F is a functional term, like gene, or wing, or belief, or clock, or
prison, or money, or subsistence farming. We seek a generalization of the form

$$(x)[Fx \rightarrow Gx].$$

In effect our search for a law about Fs requires us to frame another predicate,
Gx, and determine whether it is true of all items in the extension of Fx. This
new predicate, Gx, will itself be either a structural predicate or a functional
one. Either it will pick out Gs by making mention of some physical attribute
common to them, or Gx will pick out G's by descriptions of one or another of
the effects (or just possibly the causes) that everything in the extension of Gx
has. Now, there is no point seeking a structural, physical feature all members
in the extension of Fx bear: the class of Fx's is physically heterogeneous just

because they have all been selected for their effects. It is true that we may find some structural feature shared by most or even all of the members of F. But it will be a property shared with many other things – like mass, or electrical resistance, properties which have little or no explanatory role with respect to the behavior of members of the extension of Fx. For example the exceptionless generalization that "all mammals weigh more than .0001 gram" does relate a structural property – weight, to a functional one – mammality, but this is not an interesting biological law.

Could there be a distinct functional property different from F shared by all items in the extension of the functional predicate Fx? The answer must be that the existence of such a distinct functional property is highly improbable. If Fx is a functional kind, then the members of the extension of Fx are physically diverse, owing to the blindness of selection to structure. Since they are physically different, any two F's have non-overlapping sets of effects. If there is no item common to all these non-overlapping sets of effects, selection for effects has nothing to work with. It cannot uniformly select all members of F for some further adaptation. Without such a common adaptation, there is no further function all F's share in common.

Whether functional or structural, there will be no predicate Gx that is linked in a strict law to Fx. We may conclude that any science in which kinds are individuated by causal role will have few if any exceptionless laws. So long as biology and the special sciences continue so to individuate their kinds, the absence of strict laws will constitute a limitation on science. How serious a limitation will this be?

Not much, some philosophers of science will say. There is a pretty widespread consensus that many *ceteris paribus* generalizations do have explanatory power, and that they do so because they bear nomological force – for all their exceptions and exclusions. This thesis is the lynchpin of most accounts of the integrity of what Fodor has called "the special sciences." Indeed, on this view, my argument simply reveals that biology is one such "special science"; what is more, since all the other special sciences individuate functionally, we now have an explanation of why none embody exceptionless generalizations. A defense of the nomological status and explanatory power of exception-ridden generalizations may even be extended, as Nancy Cartwright has argued, to the claim that many generalizations of the physical sciences are themselves bedecked with *ceteris paribus* clauses; accordingly, the ubiquity of such generalizations in biology is no special limitation on its scientific adequacy.

Whether or not there are non-strict laws in physics and chemistry, there is a good argument for thinking that the exception-riven generalizations – the "non-strict laws" of biology – will not be laws at all. For the existence of non-strict laws in a discipline require strict ones in that discipline or elsewhere to underwrite them. If there are no strict laws in a discipline, there can be no non-strict ones in it either.

The trouble with inexact *ceteris paribus* laws is that it is too easy to acquire them. Because their *ceteris paribus* clauses excuse disconfirming instances, we cannot easily discriminate *ceteris paribus* laws with nomological force from statements without empirical content maintained come what may and without explanatory force or predictive power. Now the difference between legitimate *ceteris paribus* laws and illegitimate ones must turn on how they are excused from disconfirmation by their exceptions. Legitimate *ceteris paribus* laws are distinguished from vacuous generalizations because the former are protected from disconfirmation through apparent exceptions by the existence of *independent* interfering factors. An interfering factor is independent when it explains phenomena distinct from those the *ceteris paribus* law is invoked to explain. This notion of an independent interferer has been explicated by Pietroski and Rey [1995] in the following terms:

A law of the form

$$(1) \quad ceteris \ paribus, (x)(Fx \rightarrow Gx)$$

is non-vacuous, if three conditions are filled: Fx and Gx don't name particular times and places, since no general law can do this; there is an interfering factor, Hx, which is distinct from Fx and when Gx doesn't co-occur with Fx, and Hx explains why; Hx also explains other occurrences which do not involve the original law $(x)(Fx \rightarrow Gx)$.

More formally, and adding some needed qualifications,

   (i) the predicates Fx and Gx are nomologically permissible.
  (ii) $(x)(Fx \rightarrow Gx$ or $(\exists H)(H$ is distinct from F and independent of F, and H explains not-Gx) or H together with $(x)(Fx \rightarrow Gx)$ explains not-Gx)
 (iii) $(x)(Fx \rightarrow Gx)$ does sometimes explain actual occurrences [i.e. the interfering factors, H, are not always present], and H sometimes explains actual occurrences [i.e. H is not invoked only when there are apparent exceptions to (1)]

So, *ceteris paribus* laws are implicitly more general than they appear. Every one of them includes a commitment to the existence of independent interferers.

This conception of *ceteris paribus* laws captures intuitions about such laws that have been expressed repeatedly in the past. But note that on this account of non-strict laws there will have to be some laws in some discipline that are strict after all. If all the laws in a discipline are *ceteris paribus* laws, then the most fundamental laws in that discipline will be *ceteris paribus* laws as well. But then there are no more fundamental laws in the discipline to explain away its independent interferers. Unless there are strict laws somewhere or other to explain the interferers in the non-strict laws of the discipline, its non-strict statements will not after all qualify as laws at all. Without such explainers, the non-strict laws will be vulnerable to the charge of illegitimacy if not irretrievable vagueness, explanatory weakness and predictive imprecision. Or they will be singular statements describing finite sets of causal sequences. They will fail the test of Pietroski and Rey's definition of a non-vacuous *ceteris paribus* law.

Of course, if the fundamental laws in a discipline are not *ceteris paribus*, but are strict laws, this problem will not arise. This I would argue is indeed the case in physics and atomic chemistry for that matter. The Schrödinger wave equation does not obtain *ceteris paribus*, nor does light travel at constant velocity *ceteris paribus*. But as we have seen, in biology, functional individuation precludes strict laws; there are no non-strict ones either. Unless of course, physical science can provide the strict laws to explain away the independent interferers of the non-strict laws of biology. Alas, as we shall see, this is not an outcome that is on the cards. But apparent generalizations which are neither strict nor non-strict laws do not have the nomological force which scientific explanation requires. They are not generalizations at all. If biological explanations in which these non-generalizations figure do explain, they do so on a different basis from that of the rest of natural science.

Of course biology could circumvent limits on the discovery of laws governing the processes it treats, if it were to forgo functional individuation. By identifying the kinds about which we theorize and predict structurally, it could avoid the multiple realization problem bequeathed by the conjunction of functional individuation and natural selection.

But this conclusion hardly constitutes methodological advice anyone is likely to follow in biology or elsewhere beyond physical science. The reason is that forgoing functional individuation is too high a price to pay for laws: the laws about structural kinds that creatures like us might uncover will be of little use in real-time prediction or intelligible explanation of phenomena under descriptions of interest to us.

What would it mean to give up functional individuation? If you think about it, most nouns in ordinary language are functional; in part this preponderance is revealed by the fact that you can verb almost any noun these days. And for reasons already canvassed, most terms that refer to unobservables are functional as well, or at least pick out their referents by their observable effects. What is more to the point, the preponderance of functional vocabulary reflects a very heavy dose of anthropomorphism, or at least human interests. It's not just effects, or even selected effects which our vocabulary reflects but selected effects important to us either because we can detect them unaided and/or because we can make use of them to aid our survival. We cannot forgo functional language and still do much biology about phenomena above the level of the gene and protein. 'Plants', 'animal', 'heart', 'valve', 'cell', 'membrane', 'vacuole' – these are all functional notions. Indeed, 'gene' is a functional notion. To surrender functional individuation is to surrender biology altogether in favor of organic chemistry.

To be well established, our explanatory theories have a track record of contributing to predictive success. It is a fact about the history of biology that contributions to predictive success – particularly the prediction of new phenomena, as opposed merely to prediction of particular events, are almost entirely unknown outside of the most fundamental compartments of molecular biology, compartments where functional individuation is limited and nature has less scope to accomplish the same end by two or more different means. The startling predictions of new phenomena that have secured theoretical credibility, for example, the prediction that DNA bases code for proteins, emerge from a structural theory par excellence – Watson and Crick's account of the chemical structure of DNA. Even where relatively well developed theory has inspired prediction of hitherto undetected phenomena, these predictions, as noted above, have characteristically been disconfirmed beyond a limited range. Beyond molecular biology, the explanatory achievements of biological theory are not accompanied by a track record of contributions to predictive success, with one exception.

The generalizations in biology which have consistently led to the discovery of new phenomena are those embodied in the theory of natural selection itself. Though sometimes stigmatized as Panglossian in its commitment to adaptation, evolutionary theory's insistence on the universal relentlessness of selection in shaping effects into adaptations, has led repeatedly to the discovery of remarkable and unexpected phenomena. This should be no surprise, since the theory of natural selection embodies the only set of laws – strict or non-strict – to be discovered in biology. It is because they are laws that they figure in my empirical explanation of why there are no (other) laws in the discipline.

Continuing to credit biological theory with explanatory power in the absence of predictive power of course is tantamount to surrendering traditional empiricist strictures on scientific adequacy. For we must surrender the ideal of theory-driven improvements in predictive performance that are themselves explained by more and more general theories. By and large, philosophers of biology have accepted that generalizations connecting functional kinds are not laws of the sort we are familiar with from physical science. But instead of going on to rethink the nature of biology, this conclusion has led them to try to redefine the concept of scientific law to accommodate the sort of non-nomological generalizations that biological explanations do in fact appeal to. But philosophers who have done this have not recognized the collateral obligation to provide an entirely new account of the nature of explanation and its relation to prediction required by this redefinition of scientific law.

Elliot Sober exemplifies this approach to redefining scientific law with the greatest candor. He writes:

> Are there general laws in biology? Although some philosophers have said no, I want to point out that there are many interesting if/then generalizations afoot in evolutionary theory.
>
> Biologists don't usually call them laws; models is the preferred term. When biologists specify a model of a given kind of process, they describe the rules by which a system of a given kind changes. Models have the characteristic if/then format we associate with scientific laws . . . they do not say when or where or how often those conditions are satisfied. [Sober, 1993, p. 15]

Sober provides an example: "R. A. Fisher described a set of assumptions that entail that the sex ratio in a population should evolve to 1:1 and stay there. . . . Fisher's elegant model is mathematically correct. If there is life in distant galaxies that satisfies his starting assumptions, then a 1:1 sex ratio must evolve. Like Newton's universal law of gravitation, Fisher's model is not limited in its application to any particular place or time" [Sober, 1993, p. 16]. True enough, but unlike Newton's inverse square law, Fisher's model is a mathematical truth, as Sober himself recognizes:

> Are these statements [the general if/then statements] that models of evolutionary processes provide empirical? In physics, general laws such as Newton's law of gravitation, and the Special Theory of Relativity are empirical. In contrast, many of the general laws in evolutionary biology (the if/then statements provided by mathematical models) seem to be nonempirical. That is, *once an evolutionary model is stated carefully, it often turns out to be a (non-empirical) mathematical truth.* I argued this point with respect to Fisher's sex ratio argument in sec. 1.5. [Sober, 1993, p. 71]

If the generalizations of biology are limited to mathematical truths, then there are indeed few laws in this science. Sober recognizes this fact:

If we use the word tautology loosely (so that it encompasses mathematical truths), then many of the generalizations in evolutionary theory are tautologies. What is more we have found a difference between biology and physics. Physical laws are often empirical, but general models in evolutionary theory typically are not. [Sober, 1993, p. 72]

As Abraham Lincoln once said, calling a dog's tail a leg does not make it one. Sober provides little reason to identify as laws the models which characterize much of theoretical biology. But he is right to highlight their importance. Many other philosophers of biology have dwelt on the centrality which models play in the explanations biologists offer and in the limited number of predictions they make. Indeed, in the philosophy of biology the semantic approach to theories, which treats theories as nothing more than sets of models of the very sort Sober has described, is almost a universal orthodoxy. [See for example, Beatty, 1981; Lloyd, 1993; and Thompson, 1988.] Most exponents of this approach to the nature of biological theorizing admit openly that on this conception, the biologist is not out to uncover laws of nature. Thus, Beatty writes, "On the semantic view, a theory is not comprised of laws of nature. Rather a theory is just the specification of a kind of system – more a definition than an empirical claim" [Beatty, 1981, p. 410]. Models do not state empirical regularities, do not describe the behavior of phenomena; rather they define a system. Here Beatty follows Richard Lewontin: In biology, "theory should not be an attempt to say how the world is. Rather, it is an attempt to construct the logical relations that arise from various assumptions about the world. It is an 'as if' set of conditional statements" [Lewontin, R., 1980]. This is a view that has been expressed before, by instrumentalist philosophers of science. Despite its known limitations as a general account of scientific theorizing, we may now have reason to believe that it is more adequate a conception of biological theorizing than of theorizing in physical science. At least we can see why in biology, it may be impossible to say how the world is nomologically speaking, and it may be important to seek a second best alternative to this end. [I argue for this view more extensively in Rosenberg, 1994.]

Unlike physical science, where models are presumed to be way-stations towards physical truths about the way the world works, there can be no such expectation in biology. Models of varying degrees of accuracy for limited ranges of phenomena are the most we can hope for. A sequence of models cannot expect to move towards complete predictive accuracy, not even explanatory unification.

Consider the set of models that characterize population biology – models which begin with a simple two locus model that reflects Mendel's "laws" of independent assortment and segregation. The sequence of models must be continually complicated, because Mendel's two laws are so riddled with exceptions that it isn't worth revising them to accommodate their exceptions. These models introduce more and more loci, probabilities, recombination rates, mutation rates, population size, etc. And some of these models are extremely general – providing a handle on a wide range of different empirical cases, some are highly specific – and enable us to deal with only a very specific organism in one ecosystem. But what could the theory which underlies and systematizes these Mendelian models be like? First of all, suppose that since the models' predicates are all functional, the theory will be expressed in functional terms as well. But we know already that any theory so expressed will itself not provide the kind of exceptionless generalizations that would systematize the models in question – that will explain when they obtain and when they do not obtain. For this theory about functionally described objects itself will be exception-ridden for the same reasons the models of functionally described objects are.

Could a theory expressed in non-functional vocabulary systematize these models, explain when they work and when they don't? Among contemporary philosophers of biology and biologists, surprisingly enough, the answer to this question is no, no such theory is possible. Yet these philosophers and biologists do not recognize that this answer commits them to grave limits on biological knowledge, limits imposed by our cognitive powers and interests. That is, if, as is widely held, there are no more fundamental but non-biological explanations for biological processes, then biological explanations employing functional generalizations and mathematical models really do have autonomous explanatory force. But this can only be because biology is far more a reflection of our interests and powers than physical science.

The widely held autonomy of biological explanations rests on a thesis of *physicalist* antireductionism which has become almost consensus in the philosophy of biology. The explanations which functional biology provides are not reducible to physical explanations, even though biological systems are nothing but physical systems. The consensus physicalist antireductionism relies on two principles:

(1) The principle of autonomy of biological kinds. The kinds identified in biology are real and irreducible because they reflect the existence of explanations that are autonomous from theories in physical science.

(2) The principle of explanatory ultimacy: At least sometimes processes at the biological or functional level provide the best explanation for other

biological processes, and in particular better ones than any physical explanation could provide.

In the next chapter we will see that these two principles can only be defended by adopting views about causation in nature and about the reliance of explanatory adequacy on our cognitive and computational powers. As such, they provide support for the thesis of physicalist antireductionism at too high a price: for they commit the physicalist antireductionist to the sort of instrumental view about the nature of biological theory which these theorists explicitly reject. According to philosophers like Sober and Kitcher, it's not just that our feeble minds require a distinctively biological level of description because we cannot cope with physical descriptions in real time prediction and intelligible explanation. Their claim is much stronger. It is one about how the world is arranged. There really are autonomous causally efficacious levels of organization in nature distinctively different from and irreducible to the level of organic chemistry.

The argument for the claim that physicalist antireductionism commits one to the instrumental character of biological theory and biological explanation is to be found in section 4 of the next chapter. It goes without saying that most philosophers of biology will be unhappy with this relativization of biology to our cognitive limits and interests. They will argue that this conclusion simply reflects a fundamental misconception of biology on the inappropriate model of physical science. According to some of these philosophers biology differs from chemistry and physics not because of its limits but because of its aims. Unlike physics or chemistry, biology is a historical discipline, one which embodies neither laws nor aims at predictions any more than human history does. This approach stems in part from Darwin's own conception of the theory of natural selection as in large measure a historical account of events on this planet, and therefore not just a body of universal generalizations if true, true everywhere and always. Contemporary versions of this approach surrender Darwinian pretensions to nomological status for the theory of natural selection, and defend its epistemic integrity as a historical account of the diversity and adaptedness of contemporary flora and fauna. Again, interestingly, the outstanding exponent of this view is Elliot Sober. Following T. Goudge [Goudge, 1961], Sober insists that Darwin's theory is not a body of general laws but a claim about events on and in the vicinity of the Earth:

> The two main propositions in Darwin's theory of evolution are both *historical hypotheses.* . . . The ideas that all life is related and that natural selection is the principal cause of life's diversity are claims about a particular object

68

(terrestrial life) and about how it came to exhibit its present characteristics. [Sober, 1993, p. 7]

Moreover, Sober is committed to the claim, originally advanced by Dobzhansky, that "nothing in biology makes sense except in the light of evolution" [Dobzhansky, 1973]. These two commitments generate a thoroughly historical conception of all of biology:

> Evolutionary theory is related to the rest of biology in the way the study of history is related to much of the social sciences. Economists and sociologists are interested in describing how a given society currently works. For example, they might study the post–World War II United States. Social scientists will show how causes and effects are related within the society. But certain facts about that society – for instance its configuration right after World War II – will be taken as given. The historian focuses on these elements and traces them further into the past.
>
> Different social sciences often describe their objects on different scales. Individual psychology connects causes and effects that exist within an organism's own life span. Sociology and economics encompass longer reaches of time. And history often works in an even larger time frame. This intellectual division of labor is not entirely dissimilar to that found among physiology, ecology, and evolutionary theory. [Sober, 1993, p. 7]

So, evolutionary theory is to the rest of biology as history is to the social sciences. History is required for complete understanding in biology because biological theories can only provide an account of processes within time periods of varying lengths, and not across several or all time periods. Why might this be true? This thesis will be true where the fundamental generalizations of a discipline are restricted in their force to a limited time period. History links these limited-time disciplines by tracing out the conditions that make for changes in the generalizations operative at each period. Some "historicist" philosophers of history argue, like Marx or Spengler, for time-limits on explanatory generalizations by appeal to an ineluctable succession of historical epochs, each of which is causally important for epochs which follow, no matter how long after they have occurred. Others have held that the best we can do is uncover epoch-limited generalizations, and identify the historical incidents which bring them into and withdraw their force.

As a historical science, the absence of exceptionless generalizations in biology will be no more surprising than is their absence in history. For such generalizations seek to transcend temporal limits, and it is only within these limits, apparently, that historical generalizations are possible.

Leaving aside the broad and controversial questions about this sort of historicism in general, it is clear that many social scientists, especially sociological and economic theorists, will deny that the fundamental theories of their disciplines are related to history in the way Sober describes. Consider economic theory. Though economic historians are interested in describing and explaining particular economic phenomena in the past, the economic theory they employ begins with generalizations about rational choice which they treat as truths about human action everywhere and always. The theory is supposed to obtain in postwar America, and in seventh-century Java. Similarly, sociological theory seeks to identify universal social forces that explain commonalities among societies due to adaptation, and differences between them owing to adaptation to historically different local geographical, meteorological, agricultural, and other conditions.

For these sociological or economic theorists, the bearing of history is the reverse of Sober's picture. These disciplines claim, like physics, to identify fundamental explanatory laws; history *applies* them to explain particular events which may test these theories.

Unlike Sober's view of the matter, on this view social science is more fundamental and history is derivative. And if there is an analogy to biology it is that evolutionary theory is to the rest of biology as fundamental social theory is to history. This means that Sober is right about the rest of biology, but for the wrong reasons. Biology is a historical discipline, but not exactly because evolutionary theory is world-historical, and the rest of biology is temporally limited history. This is *almost right*. The rest of biology *is* temporally limited history. But the main principles of Darwin's theory are not historical hypotheses. They are the only trans-temporally exceptionless laws of biology. And it is the application of these laws to initial conditions that generates the functional kinds which make *the rest* of biology implicitly historical.

Evolutionary theory describes a mechanism – blind variation and natural selection – that operates everywhere and always throughout the universe. For evolution occurs whenever tokens of matter have become complex enough to foster their own replication so that selection for effects can take hold. And its force even dictates the functional individuations which terrestrial biologists employ and which thereby limit the explanatory and predictive power and thus the scientific adequacy of their discipline.

In our little corner of the universe the ubiquitous process of selection for effects presumably began with hydrocarbons, nucleic and amino acids. That local fact explains the character of most of the subdisciplines of biology. Their explanations are "historically" limited by the initial distribution of matter on the Earth, and the levels of organization into which it has assembled itself.

So, their generalizations are increasingly riddled with exceptions as evolution proceeds through time.

Thus, biology is a historical discipline because its detailed content is driven by variation and selection operating on initial conditions provided in the history of this planet, combined with our interest in functional individuation. This explains why biology cannot approach very closely to traditional empiricist standards on scientific adequacy. Its apparent generalizations are really spatiotemporally restricted statements about trends and the co-occurrence of finite sets of events, states and processes. And there are no other generalizations about biological systems to be uncovered, at least none to be had that connect kinds under biological – that is, functional – descriptions. This explains why biology falls so far short of scientific adequacy. Its lowest level laws will be the laws of the theory of natural selection. But these laws are too general, too abstract to fulfill our technological, agricultural, medical and other biological needs. What will fill our needs are not laws but useful instruments couched in the functional language that suits these interests and our abilities to deal with complexity. The puzzle that these limits on biology leave us with is what sort of understanding it is that these useful instruments – couched in the language of functional biology – can provide us with.

NOTE

Revised and reprinted from G. Masses (ed.), *Science at Century's End: Philosophical Questions on the Progress of Science* (Pittsburgh: University of Pittsburgh Press, 1999), pp. 241–252.

4

Reductionism Redux

Computing the Embryo

1 THE CONSENSUS ANTIREDUCTIONIST POSITION
IN THE PHILOSOPHY OF BIOLOGY

The consensus antireductionist position in the philosophy of biology begins with a close study of the relationship of classical genetics (Mendelism and its successors), to the molecular biology of the nucleic acids, and their immediate protein products. This study reveals that there are in fact no laws of Mendelian genetics to be reduced to laws of molecular biology, and no distinctive laws in molecular biology to reduce laws of Mendelian genetics, that the kind terms of the two theories cannot be linked in general statements of manageable length that would systematically connect the two bodies of theory; and that nevertheless, biologists continue to accord explanatory power to Mendelian genetics, while accepting that Mendelian genes and their properties are "nothing but" nucleic acids and their properties.

The first three of these observations serve to completely undermine the thesis once held in the philosophy of biology that Mendelian genetics smoothly reduces to molecular genetics in accordance with some revision of the postpositivist account of reduction. The last two observations have been joined together as "physicalist antireductionism" – so called because it attempts to reconcile physicalism – the thesis that biological systems are nothing but physical systems, with antireductionism – the thesis that the complete truth about biological systems cannot be told in terms of physical science alone. In particular, biologists and philosophers who embrace physicalist antireductionism adopt two theses about the autonomy from molecular biology of non-molecular biology ("functional" biology I shall sometimes call it – functional because it identifies biological structures and systems by their causal roles, usually their adaptationally selected effects):

(1) The principle of autonomous reality: The levels, units, kinds identified in functional biology are real and irreducible because they reflect the existence of objective explanatory generalizations that are autonomous from those of molecular biology.

(2) The principle of explanatory primacy: At least sometimes processes at the functional level provide the best explanation for processes at the molecular level.

We can find these two theses of physicalist antireductionism hard at work in the thought of philosophers like Philip Kitcher (1984) and Elliot Sober (1993), and biologists as different as Richard Lewontin (1982) and Ernst Mayr (1982).

The argument for these theses proceeds by example. Mendel's laws tell us that genes come in pairs and that only one of each pair is transmitted to each offspring. Since it is well established that Mendel's laws are not reducible to molecular ones, in accordance with principle (1), the Mendelian gene is an autonomous kind. Principle (2) is vindicated because Mendelian segregation is most fully explained by considerations from cellular physiology about the movement of chromosomes at meiosis, and not by providing the endless details of what molecule does what in the biochemical transactions that underlie meiosis. The cellular physiology of meiosis is the "right" level at which Mendelian regularities are most fully explained because explanations that begin at more disaggregated levels, like the macromolecular, would miss connections and similarities, would forgo significant generalizations, and would introduce unneeded, perhaps even irrelevant details. If meiosis is like other biological processes it does not result from a small set of molecular processes whose members are individually necessary and jointly sufficient for the assortment and segregation of genes. Moreover, many of the molecular processes implicated in meiosis also underlie other very different functional phenomena. So, appealing to the set of processes that doubtless happens to be necessary and sufficient in the circumstances for any one individual case of meiosis may be inappropriate for the next case. Seeking these sets will blind us to what all or most cases of meiosis share in common. It is what they share in common – that they are all cases of meiosis – which explains Mendelian assortment and segregation. So the argument for (1) and (2) goes.

But if molecular biology does not reduce classical genetics, how is it related to this theory? Positive accounts of the relationship are not a part of the antireductionist consensus. Pretty clearly, the nucleic acids, and their protein-products, provide the structure for assemblages cell physiologists observe.

And they provide the processes that implement the behavior of these cellular assemblages. They make the gene "concrete" by providing the underlying material that preserves and transmits the information classical genetics deals with. Molecular information about the location and structure of the genetic material also helps the classical geneticist understand where Mendel's "laws" go wrong, and what exceptions to these rules of thumb are to be expected. Molecular biology substantiated classical hypotheses about, for example, "mutation," which appear to be *ad hoc* in the absence of independent structural evidence.[1] Molecular biology also enabled the classical geneticist to apply methods and theories that had been shaped with an eye to observable phenotypes – properties of organisms – to molecularly characterized properties of molecular assemblages. Since at the level of molecular traits the Mendelian regularities are much closer to exceptionlessness, classical genetics came to be vindicated at the level of the macromolecule – where there seems to be one trait for each gene, even as molecular biology was explaining the weaknesses of Mendel at the level of the organism.

Of course another thing molecular biology did *to*, and not *for* classical genetics, was gravely to undermine its ontology. Molecular genetics reveals that there is no one single kind of thing that in fact does what classical genetics tells us (classical) genes do.[2] In this respect of course molecular genetics replaces Mendelian classical genetics. The classical theory retains a place in pedagogy for the same reason Newtonian mechanics does. A glance at physics textbooks shows that Newton's theory is heuristically useful and seriously misleading. Similarly, molecular biology shows why classical genetics is a useful instrument, even pedagogically indispensable, but is fundamentally flawed. One reason it is difficult for antireductionists to give a positive account of what molecular biology does for classical genetics is that the conclusion that classical genetics is merely a heuristic device undercuts the explanatory autonomy antireductionists wish to accord classical genetics.[3]

The philosophical consensus reflected in (1) and (2) is not substantiated in significant domains of biology. In particular, developmental biology, one of the subdisciplines which molecular biology made its focus in the two decades after its conquest of genetics, seems clearly to repudiate the downward direction of explanation countenanced by principle (2) above. Moreover, in developmental biology at least there are no deep explanatory generalizations we would miss were we to eschew the ontology of functional biology. For there are no explanatory generalizations at higher levels of organization. There are only descriptive regularities about "almost invariable" sequences at uniform locations like the "imagal disk" reflecting teleological mysteries of development which can only be explained in molecular terms. At most the non-molecular generalizations set out tasks for developmental explanation, and never provide explanations. Thus, principle (1) is without application in this compartment of biology. Even if biology's functional kinds are perfectly genuine, they lack any very satisfying or deep autonomous explanatory role in developmental biology. What is more, the success of molecular developmental biology's program suggests that, in developmental biology at least, physicalist antireductionism is little different from the emergentist dualism between the living and the non-living that embryology cast off even before it was invaded by molecular methods.

What follows if developmental molecular biology does not substantiate physicalist antireductionism? One possible conclusion to draw is that the relation between molecular biology and classical genetics is fundamentally different from the molecular biology/developmental biology relation. Perhaps "interfield" relations among subdisciplines in biology are more varied than philosophers have supposed and no interesting general theses in this area are tenable. Alternatively we may decide to reexamine the classical genetics/molecular biology interface, seeking a way to reconcile it with the relation we uncover in the case of molecular developmental biology. One thing we cannot do is rest complacent with blanket physicalist antireductionism as exemplified in principles (1) and (2).

In the next section I sketch recent work in developmental molecular biology which is sharply at variance with principles (1) and (2). This work suggests that non-molecular generalizations about typical cellular embryology have little explanatory power, even when they have few exceptions. They provide the explananda for this subdiscipline, never the explanans. Section 3 explores the nature of inter-theoretical explanatory relations in this subdiscipline of biology. It sketches a way of dividing the explanatory tasks between evolutionary and developmental biology on the basis of the appropriate range of application for two apparently conflicting approaches to bio-function – those of Nagel (as developed by Cummins) and of Wright. Section 4 offers an argument that vindicates molecular developmental biologists' repudiation of physicalist antireductionism, at least for their part of the discipline. It shows that principles (1) and (2) can only be defended against macromolecular explanations throughout biology by a weakening that deprives them of the ontological strength required to bear the weight of physicalist antireductionism.

2   DIFFUSIBLE MORPHOGENS: FROM DORMATIVE VIRTUE
TO THE HOMEOBOX

In *The Structure of Biological Science* I wrote that

> Nothing is more striking in biology than the apparently goal-directed phenom-
> ena of embryology and development . . .
>
> It has long been assumed that descriptions and explanations of goal-directed
> systems were ultimately to be cashed in for nonteleological theoretical expla-
> nation at the level of molecular biology. Developmental biologists do not seem
> at present to be very close to such molecular explanations of cellular develop-
> ment, still less to such explanations of the emergence of whole organs like the
> chick's wing. [Rosenberg, 1985]

A decade later this is a description of developmental biology that has been
overtaken by events. One leading developmental biologist broaches the central question of the discipline in terms that suddenly substantiate the most
reductionistic of aspirations. Here is how Lewis Wolpert sees the state of play
in developmental biology:

> Over the past 20 years, progress in developmental biology has been so
> dramatic that developmental biologists may be excused for having the view,
> possibly an illusion, that the basic principles are understood, and that the next
> 20 years will be devoted to filling in the details. The most significant advances
> have come from the application of molecular techniques and a greatly improved
> understanding of cell biology. *So we can begin to ask questions – like whether
> the egg is computable. . . .*
>
> Will the egg be computable? That is, *given a total description of the fertilized
> egg – the total DNA sequence and the location of all proteins and RNA – could
> one predict how the embryo will develop?* This is a formidable task, for it implies
> that in computing the embryo, it may be necessary to compute the behavior of
> all the constituent cells. It may, however, be feasible if a level of complexity
> of description of cell behavior can be chosen that is adequate to account for
> development but that does not require each cell's detailed behavior to be taken
> into account. [Wolpert, 1994]

The significance of Wolpert's tentatively affirmative answer to the question
whether the embryo is computable turns on how we are to understand his use of
the term "computable." Wolpert must mean something more by "computable"
than the notion of computability of mathematical functions, which makes a
given output for a given input a matter of algorithmic, mechanically decidable
processes. For so understood the thesis that the embryo is computable from

macromolecules alone will not even be controversial among biologists, including antireductionist biologists. A mathematical function is computable if a machine can execute it. The system which builds the embryo out of macromolecules is a machine, albeit one cobbled together by natural selection. Accordingly there is a computable function that this machine implements.

To avoid triviality Wolpert's claim must be understood as making restrictions on the form of the algorithm as well as the mapping of molecular input-to-embryo output. For an algorithm that adverted to cellular mechanisms not themselves "computable" from the nucleic acids and the proteins that compose the fertilized egg would hardly vindicate the hope to predict the development of the embryo from a description merely of DNA, RNA and proteins.

Thus, emergentists, holists, indeed vitalists, could agree that the embryo is computable, provided they could pack the function that maps molecules into organisms, with their favorite downward causal forces, undisaggregatable functional units, or for that matter vital forces and entelechies.

The requirement that the function rendering the embryo computable not advert indispensably to factors beyond the macromolecule is implied by Wolpert's admission that "it may be necessary to compute the behavior of all the constituent cells" of the embryo. Presumably "computing" the behavior and properties of cells means predicting this behavior and these properties from a description of the nucleic acids and proteins constituting the fertilized egg. The cell is to be at most a way station on the path to the developed organism, one algorithmically dependent on macromolecules alone. But, Wolpert's thesis must be understood as claiming that the function which renders the embryo computable will take us from macromolecules to organisms without having to pass through the way station of a complete description of all the constituent cells, or even any of them. And even if the cellular way stations are necessary, it may be sufficient to compute only some subset of cellular behavior.

The function Wolpert needs will have to be in a sense "decomposable" as well as computable: complex components of the function – say components describing cellular mechanisms – will have ultimately to be transparently decomposed into simpler components that do not invoke cellular machinery.

There is a further attribution we need to make to Wolpert's hypothesis in order to reflect the significance attached to it by developmental molecular biologists. It's not just that the function is computable, and decomposable, but that it enables us to explain the vast diversity of actual and possible morphologies on the basis of a quite limited stock of elements – maternal proteins, genes, and rules for combining them. Computationalism is overwhelmingly attractive in cognitive science because it enables us to explain the power to encode and decode an indefinitely large class of signals on the basis of a finite stock of recognizable elements and composition rules. For similar reasons, the computationalist in developmental molecular biology will hold that, unless one condition obtains, we can surrender all hope of any completeness and generality in understanding how diversity in development is possible, let alone actual: the vast diversity of form is explainable from a tractable base of a relatively small number of regulatory and structural genes (and their protein products) combined by a similarly small number of combination rules.

It is in this sense, then, that we should understand Wolpert's hypothesis that the embryo is a computable function: the function is decidable, decomposable and relatively simple.

In order to show what Wolpert's optimism, that the function satisfies these conditions, bodes for contemporary accounts of reductionism in biology, it would be helpful to sketch some of the results that sustain this optimism. In the late sixties Wolpert was already attempting to construct a theory which might guide the search for a purely molecular account of development [Wolpert, 1969]. The chemical mechanism which Wolpert suggested involved a "diffusible morphogen" – a chemical whose concentration gradient would decline with the distance from its source in the embryo, and would switch on different developmental patterns in different parts of the embryo depending on this concentration and the sensitivity of molecular receptors on cell surfaces or within them. The theory was advanced as the simplest mechanism to explain certain striking experiments, but there was no independent evidence for the existence of such a substance. At the start, the notion of a "diffusible morphogen" had all the empirical content of Voltaire's "dormitive virtue." This is where molecular biology enters the story.

Wolpert's model system is the chick limb, but the story is more simply and dramatically told for the fruit fly, *Drosophila melanogaster*.

The fruit fly embryo begins as a single fertilized egg and within 24 hours emerges as a larva, which then passes through three molts, until it pupates, and emerges 9 days later as an adult. The head, thorax, wings, legs, abdomen all develop from segments which are already clearly differentiated in the first few hours after fertilization. Developmental biologists have long been able to trace out these steps, and even to work back to the earliest events in the egg after fertilization, including the striking multiplication-division of nuclei within the single cell, after which each nucleus is enclosed in a cell of its own. By observation of some cells, their individual developmental fates can be mapped. Little in the generalizations describing this process well known to *Drosophila* biologists is more than description of fairly regular sequences. Little seems explanatory. Indeed, the sequences traditional developmental biology reports express the deep mystery of teleology at its richest.

The developmental molecular biologist's task is to "discharge" this teleology (in the way the cognitive psychologist is expected to discharge intentional homunculi). For example, for the developmental molecular biologist it is no explanation of the rapid synchronous division of the nuclei before cellularization to point out that the genetic information they contain *is needed to* direct subsequent differentiation throughout the embryo. This is evolutionary or ultimate biology. Developmental molecular biology is not satisfied with ultimate-adaptational explanations; it seeks proximal ones.

In 1995 the Nobel Prize in Medicine honored the molecular explanation of the process whereby the fruit fly embryo becomes the larva.[4] It is now known that the development of the embryo of the fruit fly is the result of a cascade of Wolpert's diffusible morphogens, and over the last decade or so they have been identified, and the genes that express them characterized. Development of the egg into an embryo and eventually a fly requires initial differentiation between the back (dorsal) and front (ventral) surfaces, and the front (anterior) and back (posterior) of the animal. Both differentiations are the result of a chemical gradient in the concentration of a protein that regulates gene expression by binding to sites of the DNA that turn on and turn off genes which produce still other proteins. Let's consider how anterior/posterior differentiation proceeds.

Prior to fertilization, so-called nurse cells surrounding the oocyte express the messenger RNA (mRNA) for the protein product of the *Bicoid* gene, which binds to a molecule on the oocyte's anterior end. Once fertilization occurs, the mRNA is translated into a protein which diffuses from higher concentration in the oocyte close to the nurse cells to lower concentration in the other end. This protein is a long-range "signal," which turns on one gene in high concentrations, turns it off at lower concentrations, and whose absence turns on still a third gene. The bicoid gene expresses a protein which turns on and/or turns off some twenty-five segmentation genes, of which there are three known groups: gap genes, whose products produce the basic (para) segments of the embryo; the so called pair-rule genes – a brace of genes whose differential expression divides (para)segments into further segments; and segment-polarity genes, which orient segments. Most of the segmentation genes are known to code for further regulatory proteins, chemicals that switch on and off still other genes. And these genes are themselves all switched on by different concentrations of the product of the bicoid gene. In turn the gap genes produce proteins that spread out as diffusible morphogens, constituting positional signals switching on and off genes in nearby cells and controlling their development. The next level of spatial differentiation is controlled by the pair-rule genes. These two appear to code for proteins that diffuse to transcription sites in the genomes of neighboring cells, switching on other genes, and so on until the production of regulatory proteins gives way to that of structural proteins in amounts that result in the different cells, tissues, and organs of the fly. As yet much less is known about these structural proteins, but it is already clear that the difference between organisms is largely a result of differences in regulatory gene products and not structural gene products.[5]

The most striking discoveries in developmental molecular biology are those which have identified the so-called Homeotic selector genes, the next level in the cascade of morphogen-producing genes after the segmentation genes. There is in fact fairly startling evidence that these genes produce the most complex of organs in a straight-line manner similar to the fixing of the fates of segments of the fruit fly embryo. Walter Gehring has reported experiments in which a previously identified homeotic gene, *Eyeless*, when activated in somatic cells all over the body of adult *Drosophila*, results in the growth of complete eyes. These eyes, including cornea, pseudocone, cone cells, primary, secondary, and tertiary pigment cells, are functional at least to the extent that their photoreceptor cells respond to light (Gehring, 1995). Gehring's team has induced eyes in the wings, antennae, halteres, and in all six legs, and they were able to do so in 100% of the flies treated under conditions in which the *Eyeless* promoter gene functions.

Eyeless appears to be a "master-control" gene whose activation by itself is necessary and sufficient [Gehring, 1995, p. 1791] to trigger a cascade of genes harbored in all the cells, but normally silent in all but those which give rise to eyes. Presumably, a protein coded by *Eyeless* binds to some set of genes, switching them on and producing a cascade of proteins that ectopically builds an eye on the fly's back, or under its wing, or on its haltere, or even on the end of one of its antenna. And this set of genes is of course to be found in every nucleus in the fruit fly's body. Gehring estimates that the number of genes required for eye-morphogenesis is 2500 (out of approximately 17,000 genes in the *Drosophila* genome), and that all are under direct or indirect control of *Eyeless*. Moreover, *Eyeless* appears to directly control later stages of eye-morphogenesis. Apparently, the same master-control gene functions repeatedly to switch on later genes, crucial to eye-development, suggesting that evolution has employed the same developmental switch several times in selecting for eye-developing mechanisms.

What is more, *Sey*, the mouse gene homologous to the fruit fly's *Eyeless* gene, will produce the same result when inserted into fruit fly somatic cells and switched on. And there is evidence that in the mouse, *Sey* is a master-control gene as well. Proteins encoded by the homologous genes in the two species share 94% sequence identity in the paired domains. Gehring's laboratory has identified counterparts to the fruit fly's *Eyeless* gene, which are implicated in eye-development across the whole range of species from planaria to squid to humans. *Eyeless* and its homologues may be present in all metazoa.

These results suggest that one of the most complex of organs is built by the switching on of a relatively small number of the same genes, across a wide variety of species, and that the great differences between, say, mammalian eyes and insect eyes are the result of a relatively small number of regulatory differences in the sequence and quantities in which the same gene products are produced by genes all relatively close together on the chromosome, and that these genes build the eye without the intervention of specialized cellular structures beyond those required for any developmental process. Identifying the other genes in the cascade that produces the entire eye should in principle be a piece of normal science, which will enable the developmental geneticist to "compute" the eye from nucleic acids and proteins alone. For if switching on *Eyeless* can create the eye, surely its creation is "computable" at least in principle. Moreover, as Gehring concludes, "The observation that mammals and insects, which have evolved separately for more than 500 million years, share the same master control gene for eye morphogenesis indicates that the genetic control mechanisms of development are much more universal that anticipated" [Gehring, 1995, p. 1792]. The more universal, the simpler as well, for a very complex package of instructions to build the eye is open to mutation, and recombination that must reduce its universality as evolution proceeds.

Once *Eyeless* is switched on by regulatory proteins, nothing else beyond the constituent macromolecules is needed, apparently, to "compute" the eye. But, comes the reply, surely, the molecular developmental biologist cannot simply build an eye, still less an animal *in vitro*, by combining the right macromolecules in the right proportions in the right sequence, in the right intervals. Surely the cellular milieu in which these reactions take place is causally indispensable and so the claim to compute the embryo is exaggerated at best.

Molecular developmental biologists may admit that building the eye *in vitro* is beyond the present dreams of the discipline, but in the long run the cellular milieu cannot serve as a "black box" through which macromolecules are transformed into embryo structures. That is the whole point of the "computability" claim. Just as cell-cell signaling is ultimately to be cashed in for a chain of molecular interactions that extend from one stretch of nucleic acids to another across several lipid bi-layers (the cell membranes), all other cellular structures implicated in the machinery of differentiation will eventually have to be disaggregated into their molecular constituents, if development is fully to be explained.

Here, in developmental molecular biology, there is no room for downward explanation, in which some regularity at the level of cell physiology plays a

81

role in illuminating the molecular processes that subserve development. And this is for two reasons. The first is that such generalizations as obtain at the level of cell physiology are either wholly descriptive histological reports or functional regularities which are developmental biology's job to explain, but never provide its explanations. This is a point to which I return below. The second reason is that cellular structures only come into existence through the molecular processes that precede them. There is in developmental molecular biology therefore no scope for claims about the indispensable role of cellular structures in these molecular processes. The future cannot cause the past.

It is true that maternal cellular structure, as well as that of the oocyte and the sperm cells, plays a role in our explanation of how the molecular process builds the embryo. One might accept that developmental molecular biology takes as given maternal cellular structures, and perhaps even assumes the cellular structure of the egg. It then aims to explain everything else that happens in embryological processes without adverting further to ineliminable cellular physiology. After all, the switching on and off of regulatory and structural genes is not by itself causally sufficient for cellular differentiation, any more than striking a match is causally sufficient for its lighting. Much else is required, and much of this – it will be held – can only be described, at least for the moment, in terms that advert to cellular structures, ones which have an explanatory role in regulatory and structural molecular pathways. Below I explore the significance of the fact that developmental molecular biologists do not seem to be interested in providing all the conditions – molecular or cellular – that would be causally sufficient for the development of the embryo.

But the idea of according cellular structure a permanent and ineliminable autonomous explanatory role along with molecular processes does not represent the program of molecular developmental biology. Developmental molecular biologists may delimit their area of interest to what happens after oocyte formation. However, it would be a double standard for them to suppose that "computability" might be vindicated for the embryo but not for its mother. Unraveling the molecular details of development has to start somewhere, and the natural starting place is the stage composed of the fertilized egg, and the cellular structure that supports fertilization. On this stage the first players to appear are maternal RNAs and proteins. But because the stage, the reproductive apparatus of the mother, is also the result of a developmental process, molecular developmental biology's commitment to "computability" means that this reproductive apparatus is ultimately to be given a molecular account. And indeed, much of the attention of molecular developmental biology is devoted to uncovering the molecular mechanisms of the differentiation of reproductive cells at the very beginning of embryological development. For these cells are among the most distinctly different and specialized from the very beginning of development.

It remains to be seen, of course, whether the program of computing the embryo can be carried out. And there are certainly developmental biologists who have expressed doubts about its ultimate success. However, some of this dissent may rest on a terminological presupposition that divides computationalists from non-computationalists in developmental molecular biology. T. J. Horder has long argued that "a number of diverse but well-known categories of data call into question the adequacy of a purely genetic view of form." He concludes:

> The available evidence (complete and selective though it may be) is incompatible with any simple conception of a one-to-one relationship between morphology and the normal genetic programming of morphology whereby each morphological locus derives its unique features through independent control by appropriate specific genes. The complexity of the evidence is such that there is no obvious way to deduce how or in what degree genes alone or in combination define and limit morphology. Since detection of genetic factors depends on morphological outcomes, the situation is, in the absence of independent information about intervening mechanisms, essentially circular. [Horder, 1989, p. 324]

One matter worth immediately addressing is the threat of circularity under which molecular developmental biology may lie. Taking it too seriously would undermine the search for all underlying unobservable processes which can initially only be individuated in terms of the phenomena the factors are invoked to explain. It is only when no means are forthcoming independent of the phenomena they are invoked to explain, that inferences to explanatory factors should be suspect as circular. This is the "dormitive virtue" problem that Voltaire so effectively identified. It might have been a serious criticism of Wolpert's diffusible morphogens: had nothing ever been shown for them – no molecules identified, no genes to express them, no assay to measure their diffusion – then the initial circularity would have turned out to be vicious. This conspicuously was not the case.

More important, Horder's doubts may in fact belie a fundamental agreement with advocates of computability. For the non-genetic factors Horder wishes to invoke as integrally involved along with the genes in determining form are apparently themselves molecular, or potentially so.

Horder writes:

> A simplistic dichotomy between "genetic" and "environmental" factors neglects the internal environment of the organism. This is not normally amenable to manipulation. . . . The possibility therefore remains open that "internal"

nongenetic factors may represent an unsuspectedly large and systematic contribution to the determination of form. [p. 323]

... The more we discover in molecular terms about genetic systems, ... the clearer it becomes why "nongenetic" considerations cannot be ignored. In each cell, differentiation depends on the selective expression of large numbers of structural genes. It is the pattern of selection of expression in different cells that defines an organism as well as the character of the structural genes themselves. The selection process can only be mediated across dispersed genetic elements by selective molecules within a cell's repertoire other than DNA. ...

The fundamental facts of cell biology make it inevitable that many factors remote from DNA are necessary for, and set limits upon, the ways in which DNA can influence morphology. [Horder, 1989, pp. 324, 325]

There is nothing in these passages with which a computationalist about the embryo should disagree. To begin with, all recognize that genes by themselves code for no phenotype without an environment to interact with, and to select from among their gene products. As the debate about genic selection revealed,[6] the environment of the gene begins with the molecular milieu of the nucleus. It is equally clear that the computability thesis is not committed to the notion that nucleic acids are enough to determine the embryo. It invokes other macromolecules, at the start maternal proteins and RNA, whose products kick off development. Moreover the diffusible morphogen is just the sort of intercellular messenger Horder requires to select developmental fates for cells that produce morphogenic patterns.

So, where does the dispute among developmental biologists lie? It lies, I think, in divergent hunches about the degree of complexity of the "function" that takes molecular inputs into embryological outputs, and the similarity of the "function" across widely divergent species. Wolpert and other reductionists hold that the absolute number of genes – regulatory and structural – that determine morphology is relatively small, and that the differences among organisms are due to relatively simple differences in the order in which regulatory genes are switched on and off.

As noted in section 2, "computability" in developmental molecular biology turns out to be a variant of "computationalism" in cognitive psychology: The thesis that a capacity to produce an indefinite variety of forms from a finite stock of units – the genes – is only explicable if the stock of units can be combined in accordance with a syntax – rules about switching on and off given in the case of development by natural selection – to produce the variety we know is possible.

Those who dissent from "computationalism" in fact have no stake in the denial that there is some sort of function from molecules to embryos, and physicalists among them will even grant that the function be in principle decomposable in the sense advanced in section 2. The dispute is whether the function is at the same time simple enough actually to be formulated and powerful enough to explain the morphological diversity nature manifests. Consider, for example, Gehring's claim that "only" 2500 genes may be involved in the most complicated of systems, the eye. If these 2500 genes interact in accordance with 2 to the 2500th power different "rules" for the construction of gene-products, then the function from molecule to eye will not be computable for any practical or even explanatory purpose.[7] Whether the computable function is like this will clearly be an empirical dispute about contingent matters of fact. But it is one with implications for the philosophy of biology and on which philosophical arguments may bear.

3 REDUCING FUNCTIONS WITHOUT REDUCTION-FUNCTIONS

Reductionism in developmental molecular biology has some modest morals for debates in the philosophy of biology, and at least one quite unmodest lesson. The principal explanatory task of molecular developmental biology is to discharge the teleology of functional developmental biology. This is not news of course. For the better part of a century now, reductionists have held that cashing in teleological for nonteleological processes is an important part of the agenda of physicalist science. What is news, however, is that the program has begun to be successfully articulated by molecular developmental biology. The promissory notes are being honored in macromolecular *specie*. But for our purposes what is significant is that because it aims to cash in physicalism's promissory notes, developmental molecular biology will not countenance downward causation in the spirit of principle (2). Nor does it advert to explanatory generalizations that make autonomous the kind terms of functional biology, as principle (1) proclaims.

Starting with a generalization like

(f)   the function of the fifth segment in the *Drosophila* is to produce the wing

molecular developmental genetics provides a nonfunctional account that explains (f). These explanations in developmental molecular biology nicely substantiate the Cummins/Nagel theory of functional explanation (Nagel, 1961; Cummins, 1975). I describe it as the Cummins-Nagel theory because, unbeknownst to almost everyone, including Cummins (who thought he was refuting Nagel), his account of function works because it relies on Nagel's directively organized system approach, as I shall illustrate by application to an example.

Explanations in molecular developmental biology proceed by showing how functional processes are implemented by macromolecular ones that operate biochemically. It cashes in functional attributions for non-functional attributions in accordance with Cummins' schema (Cummins, 1975 in Sober, 1994, p. 64):

(3) **x** functions as an **F** in **s** (or the function of **x** in **s** is to **F**) relative to an analytical account **A** of **s**'s capacities to **G** just in case **x** is capable of *F*ing in **s** and **A** accounts for **s**'s capacity to **G** by appealing to the capacity of **x** to **F** in **s**.

Instantiated in terms of the function of the product of the bicoid gene, (3) reads

(3') The bicoid gene-product functions as a morphogen in *Drosophila melanogaster* embryo (or the function of bicoid gene-product in *Drosophila* embryo is to act as a morphogen) relative to **A** – an analytical account of the *Drosophila* embryo's capacities to segment, just in case the bicoid gene-product is capable of being a morphogen in the *Drosophila* embryo and **A** accounts for the *Drosophila* embryo's capacity to segment by appealing to the capacity of bicoid gene-product to act as a morphogen in *Drosophila* embryo.

The payoff of this account is to be found in how it "discharges" the teleological attribution to the bicoid gene-product by analyzing capacities into biochemical regularities at the macromolecular level. And this is where Nagel's directively organized system approach enters, as I will show shortly.

Following Cummins, we explain functions by capacities, which are already well understood: capacities are sets of dispositions. The capacity of the *Drosophila* embryo to segment is explained by appeal to a number of other capacities of components of the *Drosophila* embryo – the regulatory genes and their products "such that programmed manifestation of the components' dispositions results in or amounts to a manifestation of the embryo's capacity" [Cummins, 1975, in Sober, 1994, p. 63]. Basic capacities or dispositions – like the disposition to bind preferentially to a certain DNA sequence – are explained by "instantiation": showing how the disposition is realized in the things which have it.

But wait, what is the role of the expression "programmed manifestation" in this analysis? By programmed manifestation, Cummins tells us, he means "organized in a way that could be specified in a program or a flow chart." Elsewhere the appeal to the notion of a program has either trivialized accounts of teleology or function (see for example, Mayr, 1982, p. 48) or wrongly required us to count non-functional processes as functional ones. When "program" is understood to itself describe a process or a state requiring intentional intervention or judgment, appeals to it are question-begging. When programs are treated as mechanically decidable procedures, a variety of purely physical processes appear to qualify as teleological. For example, changes in the values of pressure, temperature and volume in a gas can be expressed as a program with the function of maintaining their relation to the gas-constant r. But such a program is clearly no basis for identifying a gas as a functional or teleological system.

Nagel dealt with both these potential counterexamples by requiring that a goal-directed system consist in a set of subsystems, whose non-teleological behavior implements the teleology of the whole system, just as Cummins does. However, Nagel gives an account of the nature of the implementation which avoids the counterexamples. A directively organized system is one composed of subsystems which are themselves (ultimately) not directively organized but interact in such a way that a state of the whole system is maintained – either the state of tracking some goal through environmental variations, or the goal state itself – by the temporally asymmetric feed-forward and/or feed-back effects of changes in the values of causal variables of each of the subsystems on one another. This temporal asymmetry requirement precludes counterexamples like those which turn ideal gases into teleological systems. The flow chart of the program under which the subsystems interact is mechanically decidable, and engenders no regress to further teleology.

One issue on which molecular developmental biology's discharge of functions should cast light is the vexed relationship between developmental and evolutionary biology. In particular, it is for evolutionary biology to answer the question (to the extent it is answerable) of how and why individual non-teleological processes came to be packaged together in such a way as to constitute directively organized systems, and how these directively organized systems should themselves be packaged into such large systems. This individuation of packages and their components via their "bio-function" is underwritten by the theory of natural selection in the ways Larry Wright uncovered in his teleological or etiological analysis of functional attributions [Wright, 1976]. It is for evolutionary biology to individuate the entities and processes which developmental biology explains; it does so by an implicit or explicit appeal to their adaptational etiologies – their bio-functions. It is for developmental biology to explain how they accomplish these bio-functions by appeal to their capacities. Thus developmental and evolutionary subdisciplines are distinguished in part by the differing notions of function which they imply, and related by the fact that developmental biology's functional attributions are dependent on evolutionary biology's functional attributions. If the latter are wrong, then the former are ungrounded. But etiological individuation simply sets the problems for developmental biology, it is no part of their ultimate solution.

For the moment we have only the sketchiest idea of how for example the eukaryotic cell might have evolved from the packaging together of the mitochondria and bacteria. It may be possible to recover the sequence of events which packaged the various RNAs and DNA into the directively organized system which produces proteins. But we will be able to do this only if there are but a small number of possible routes from the existence of individual nucleic acid bases to macromolecules.

By providing an evolutionary etiology that explains the persistence and multiplication of these packages of non-teleological subsystems, evolutionary theory finds its way into every compartment of developmental biology, even the most fundamental level.[8] But molecular developmental biology takes these packages as given, and uncovers the molecular details of how they operate. These details both highlight evolutionary lineages in the homologies they reveal, and more importantly, enable the evolutionary molecular biologist to provide an account of how selection actually brings molecular subsystems into directively organized wholes. Above the level of the macromolecule, functionally characterized structures are supervenient on a large disjunction of alternative actual and possible mechanisms. Since each of these different mechanisms is the result of a different causal process, there is no unique path of adaptations that resulted in a functionally characterized cell, tissue, organ, structure or behavior, nor probably a manageable disjunction of them. And what these pathways all do share in common – their selectively similar effects – is just what is to be explained! At most evolutionary biologists can identify design solutions and tell "just so" stories about how they might have emerged. At the level of the macromolecule the chemical and physical constraints may be narrow enough so that a manageably small number of causal routes from mere matter in motion to a macromolecular subsystem can sometimes be uncovered, by for example laboratory experiment.

This relatively clean division of responsibilities between developmental and evolutionary biology makes it particularly clear why selection operates so overwhelmingly at the earliest stages of development of the fertilized egg. Packages that will not work together are wiped out early, and adaptations at any level above the macromolecule have no chance of even appearing unless they are the result of macromolecular combinations selected for their own relatively immediate adaptive advantage.

Another of the modest lessons of molecular developmental biology is the light it sheds on the unit of reduction. Reductionism-for-and-against is traditionally viewed as a debate about statements, theories, laws, models, or other linguistic items that reflect generalizations. But reductionism in developmental biology is not a thesis about laws, or surrogates for them like practices, models, other conceptual items. It is about specimens, or perhaps even about particular organisms. Having traced the developmental pathway from maternal messenger RNA all the way to the adult fly in one batch of fertilized eggs, the developmental molecular biologist is satisfied to find homologies and similarities in the development of other organisms, and is even willing to accommodate different pathways in the same or similar organisms with slightly different genomes or different environmental milieu. The vindication of molecular biology is more a matter of proving an existence proof for one or a small number of purely molecular pathways to a biological system. It is not a matter of tracing all or most or even many of the different pathways that eventuate in the same outcome.

Why is an existence proof that at most establishes just one complete story from molecule to animal for one sample of embryos sufficient in the developmental molecular biologists' program to sustain the "computability" thesis? Part of the answer must be that developmental molecular biologists understand that the diversity of organisms that evolution fosters is so great that empirical generalizations about development which combine strength and simplicity are unlikely. Developmental molecular biologists seem to recognize that theirs is not a science that aims at laws, but the application of other nomothetic sciences to tracing out singular causal chains. If there are generalizations in this discipline, they are either of the sort developmental biology aims to explain or else they are about laboratory techniques, methodologies, assays, tricks that will turn genes on or off, produce mutations, move genetic elements from one place to another.

The existence proofs molecular developmental biology seeks for its computable functions establish a general possibility on the basis of a small number of actualities: showing how a complex biological system does emerge from a purely chemical process, in one case, establishes the possibility that it can so emerge in many others. At this stage it is more important to go on to another model system – a more complicated one – to establish yet another general possibility. Going back and tracing out more actual routes is an exercise left to later in the research program.

The apparently attainable object of the subdiscipline is not a function that will map every molecular input to an embryological output. One mapping at most establishes the existence of a class of functions. But why should un-covering one causal chain from maternal RNA and regulatory proteins all the way to the developed embryo suffice to establish the existence of the gen-eral function? And why are developmental molecular biologists satisfied with establishing merely the existence of the function, not its full specification? These are important questions. But the very fact that they present themselves suggests that there is something quite different going on in molecular devel-opmental biology than what philosophers of biology have supposed happens when a more fundamental theory reduces, "extends," or otherwise unifies a less fundamental one.

Reductionism and its surrogates are viewed as driven by an imperative to unify. As Kitcher writes, "the unification of our account of the world is a cognitive desideratum for us, a desideratum that we place ahead of finding the literal truth on the many occasions when we idealize the phenomena. The causal structure of the world, the division of things into kinds, the ob-jective dependencies among phenomena are all generated from our efforts at organization."[9] But the reduction molecular developmental biology aims at places literal truth ahead of idealized models that unify, and indeed substitutes for unified accounts of apparently homogeneous phenomena disaggregation into distinct pathways that give the causal structure of the world. In this de-partment of biology "the objective dependencies" are not the result of our efforts at organization, they are the results of our search for the literal truth.

Why this is so is an important question. But there is a prior one that awaits the defender of reductionism in developmental molecular biology. If physicalist antireductionism is correct, there may be some developmental processes for which the function Wolpert seeks does not even exist. In that case, molecular developmental biology will have to curtail its reductionist pretensions. For all Wolpert knows, there may turn out to be obstacles to "computability" in the form of non-molecular embryological processes that honor principles (1) and (2).

4   DEVELOPMENTAL MOLECULAR BIOLOGY'S EXEMPTION FROM PHYSICALIST ANTIREDUCTIONISM

Whether the embryo is computable, whether there is a function mapping every biological outcome to a unique set of molecular inputs, is a contingent matter. This makes reductionism in developmental molecular biology an empirical theory, and not an a priori doctrine. And if physicalist antireductionism holds sway elsewhere in biology, who is to say it will not obstruct the reductionistic aspirations of developmental molecular biology.

What developmental molecular biology needs is an exemption from the writ of physicalist antireductionism. It needs an argument against principle (1) – the levels, units, kinds identified in functional biology are autonomous and irreducible, just because they figure in explanatory generalizations we would miss if we did not adopt the language of these functional kinds; contra principle (2), it needs to show that in development at least processes at the biological level never provide the best explanation for processes at the molecular level.

These two principles entail a commitment to "downward causation," to the thesis that functional states of biological systems cause molecular processes, and do not do so simply owing to their molecular composition and properties. Now, this is a thesis that some antireductionists willingly embrace. But downward causation is a commitment intolerable to developmental molecular biology, or at least to Wolpert's program. If molecular developmental biology's commitment to physicalism is inimical to the very possibility of downward causation, then it will be exempt from the writ of physicalist antireductionism.

It is relatively easy to see why principles (1) and (2) require downward causation. Principle (1) is grounded on the notion that to be real, and not just artificial, a kind or category's instances must have distinct causal powers, ones reported in causal laws we would miss if we did not commit ourselves to their reality. But the causal powers of biological kinds must be distinct from the causal powers of assemblages of molecular properties. Otherwise, the distinctness of biological kinds from molecular ones would be threatened. For their causal powers would not be distinct from those of molecular kinds.

Principle (2) tells us that the instantiation of these non-molecular real properties of biological systems sometimes provides the best, most complete explanation of the instantiation of molecular properties. Now, since these explanations are objectively the best ones, and not just the most heuristically tractable for creatures of our cognitive and computational powers, their explanatory excellence presumably reflects the accuracy of their reports of causal processes. Consequently, in these cases the direction of best explanation will follow the direction of causation: i.e., if explanation is downward, then this must be owing to the downward direction of causation from the level of the functional to the level of the molecular.

Keep in mind that the downward causation required here is not one that can be cashed in for more fundamental "upward causation" from molecular kinds to biological ones. It cannot be the case that biological processes cause molecular ones just in virtue of the biological processes *being* themselves molecular. For this would turn downward causation into a mere way station for more upward or sideways causation.

Now, physicalism holds that all biological properties are realized by combinations – sometimes vastly complex combinations – of molecular properties: Whenever an organism or system instantiates a biological property F, it has some molecular property M such that M realizes F in systems or organisms that have F. Of course, the agenda of developmental molecular biology is explicitly to uncover how molecular properties realize biological ones. The notion that functionally described processes are composed of nothing but macromolecular processes is a core commitment all physicalist biologists share.

To see the problem downward causation makes for physicalism, suppose that a given functional property, say segmentation by a parasegment in *Drosophila*, has a macromolecular effect, say blocking the diffusion of a morphogen. If segmentation supervenes on some complex conjunctive/disjunctive macromolecular property, Mx, then every instance of segmentation is realized by some instance of the molecular property Mx. Downward causation from the biological to the molecular requires, for example, that sometimes cellular ion blocks chemical diffusion of the morphogenic molecule. But of course, segmentation is implemented molecularly by the parasegment, because the parasegment instantiates the (very complex conjunctive/disjunctive) macromolecular property M. If the complex macromolecular property M is how the parasegment implements segmentation, then the question arises why it is that the parasegment blocks the diffusion of the molecule by segmenting, and not the macromolecules that implement the segmenting which do the molecular-blocking job? Surely this is not a case of overdetermination by independent molecular and cellular processes. We need to block the claim that the macromolecular property M does the job, just because we attributed the cause of morphogen blocking to cell segmentation – this is its distinctive causal power. We cannot also attribute this blocking power to segmentation's molecular implementation without depriving segmentation of the distinctive causal role which guarantees its autonomy, indispensability and irreducibility. But this means, contra physicalism, that the segmentation has distinct non-physical causal powers uncovered in functional biology.

Physicalism can be reconciled with downward causation, but only at a cost that is probably too heavy for developmental molecular biology and certainly prohibitive for most physicalists. The problem is to combine two claims: the claim of mereological – whole/part, "nothing but" – dependence of concrete biological systems on their constituent macromolecular implementations must be combined with the claim that the biological systems have *distinctive* causal powers – powers with macromolecular effects different from the effects of the macromolecular assemblages that implement them. Remember, without distinctive causal properties, there is no basis to accord functional states, entities and processes with reality autonomous from molecular phenomena. Principles (1) and (2) derive autonomy from causal/explanatory role.

The physicalist antireductionist needs to block the shift of causal powers from the biological down to its macromolecular implementation. This requires two controversial "moves" in the philosophy of science.

First we need to embrace the view that causation is a relation among states, events and processes that is conceptually dependent on explanation. That is, the notion of explanation is more basic than the notion of causation, and the latter can only be understood on the basis of an understanding of the former. This is hard to swallow because it makes a fact apparently about the world, causation, depend on a fact about explanations we request and provide. But it is a view that at least some physicalist antireductionists embrace (see Kitcher, 1993, p. 172).

To the thesis that explanation is conceptually prior to causation we must add another controversial claim about explanation: that it is heavily "pragmatic": Explanation is pragmatic roughly when it is viewed not as a relation just between propositions, but as a relation between questions, and answers offered in a context of an interlocutor's beliefs. These beliefs reflect presuppositions of interlocutors which together determine the correctness or goodness or explanatory worth of various answers.

Here is how pragmatism about explanation combined with its conceptual priority to causation can save physicalist antireductionism. Suppose that some biological system a has some functional property F, and a's having the functional property F is implemented by a having some complex molecular property M.

So a has F because a is M, and the way F is realized in nature is through the molecular mechanism M.

Now, how can

$$a \text{ has } F$$

explain

$$a \text{ has } G \text{ (where } G \text{ is some effect of being an } F)$$

without

$$a \text{ has } M$$

(also) explaining why

$$a \text{ has } G?$$

After all, a's being an F is "nothing but" a's being an M.

The only way this can happen is when "explains" is a non-extensional relation not just between matters of fact, but between them and cognitive agents with varying beliefs. For someone who doesn't know that a's Fing is realized by a's Ming, the substitution of M for F in the explanation won't work. If we don't know about M, or understand how it does F, then a's having M doesn't explain to us what a's having F explains to us.

This means that when explanation is treated as "subjective," "pragmatic," its direction can move downward from the functional to the molecular. And if causation is just explanation (or depends on it), it too can move "downward" from the biological to the molecular, following the direction of explanation.

For cognitive agents like us many of the complex macromolecular properties on which the biological ones supervene may be too complex to uncover, to state, to employ in real-time explanation and/or prediction. Under these circumstances, the biological property may provide the best, or the only explanation (for us) of the phenomenon in question. On this view downward explanation of the sort principle (2) envisions may be possible: it may turn out that sometimes the instantiation of biological properties autonomously and indispensably explains macromolecular processes, without the functional property's instantiation being explainable (to us) by appeal to the instantiation of macromolecular properties that we are intelligent enough to uncover and use in real time.

Putting the pieces together we can now construct an account of downward causation compatible with physicalism: explanation is sometimes downward because there are contexts in which correct molecular answers to explanation-seeking questions are non-explanatory (to us). Causation will be downward in these cases if it follows the direction of explanation.

However, this reconciliation of physicalism with downward causation will not suffice for the antireductionist component of physicalist antireductionism. For antireductionism requires we interpret principles (1) and (2) on either a non-pragmatic notion of explanation, or on a conception of causation as prior to explanation.

Principle (1) holds that there are generalizations which are explanatory "objectively," not merely relative to some interlocutor's beliefs. Only such non-pragmatic "objective" explanatory power will underwrite ontological inferences to existence of functional kinds as irreducible. Principle (2) invokes a notion of "best explanation," either context free or relative to some privileged context, say, that which obtains among interlocutors with the most complete and accurate beliefs about causal processes. And this notion of "best explanation" will be circular if its presuppositions about these causal processes are explanation-dependent. For purposes of a debate between reductionists and antireductionists about principles (1) and (2) a pragmatic approach to explanation and the assimilation of causation to explanation are not merely unavailing. They are likely to undermine both principles (1) and (2) by cutting the connection between explanatory power and metaphysical commitment to what really exists independent of our interests. If biological properties are real it is because they are, like electrons, indispensable to the most well established theory under some context-free standard, or because there is good non-explanatory, observational evidence for their existence, not because they satisfy the explanatory itch of cognitive agents of our sort.

Let us accept for purposes of argument that explanation is heavily pragmatic. What no physicalist can accept is that explanation so viewed is conceptually prior to causation – i.e. that causation can be defined in whole or in part by appeal to explanation. That way lies Kantian idealism at best and social constructivism at worst. By itself however, the (pragmatically) explanatory indispensability of biological kinds will not justify their biological reality or their autonomy from the macromolecular. And if best explanations are ones which report objective causal sequences most completely, then in the end functional biology may turn out to be a way station in the direction of a thoroughly reduced and perhaps even computational molecular biology.

5   CONCLUSION

That physicalist antireductionism is the common wisdom of modern philosophy of biology might surprise the non-cognoscenti. If I am right, it should not be surprising to the cognoscenti that at least one component of modern biology does not support this consensus position. And this component – molecular developmental biology – bids fair to be at least as fruitful a source of insights into the nature of biological processes over the next few decades as molecular genetics has been in the past few. Furthermore, the commitment to physicalism it shares with the rest of biology and on which molecular developmental biology relies most directly, undermines the common wisdom of physicalist antireductionism, unless interpreted as a thesis about biologists – their interests and limits – as opposed to a thesis about the ontology and meta-physics of the biological realm.

NOTES

Reprinted from *Biology and Philosophy*, 12 (1997): 445–470, copyright © Kluwer Academic Publishers, with kind permission from Kluwer Academic Publishers.

1. As Kitcher notes, "Classical genetics makes certain presuppositions – that genes replicate, that some mutations are viable, which apparently seem impossible, given premises that classical geneticists accepted. Molecular genetics shows how these presuppositions could in fact be true, consistent with the premises classical geneticists accepted." See Kitcher, 1984.

2. As Kitcher says, molecular biology provides "a specification of the entities that belong to the extensions of predicates in the language of the earlier theory [classical genetics], with the result that the ways in which the referents of these predicates are fixed are altered in accordance with new specifications." See Kitcher, 1984, p. 364. Here Kitcher is assimilating the way in which regularities about molecular phenotypes successfully instantiate classical generalizations about phenotypes which are disconfirmed regularly at the level of gross morphological features. By substantiating these classical generalizations at the level of biochemistry, while falsifying them at the level of the cell and above, molecular biology drastically shifts causal roles away from the classical gene and towards so many different molecules as to extirpate the entire gene concept.

3. For a broader discussion of the relations between classical and molecular genetics, and their implications for intertheoretical relations generally, see Rosenberg, 1994, Shaffner, 1993, and Waters, 1990.

4. This work was carried on by Edward Lewis of Cal Tech, Eric Wieschaus of Princeton, and Christiane Nuesslein-Volaed of the Max Planck Institute.

5. A highly accessible account of the last decade and more's discoveries in the embryology of the fruit fly are recounted in Lawrence, 1992.

6. Here I am indebted to Peter H. Schwartz.

7. Note that changes in the values of pressure, temperature and volume of a gas fix one another's values instantaneously.

8. This should be no surprise. Consider the difference between DNA and RNA. The latter differs chemically from the former only because of the differences in the functions they perform. DNA stores and transmits information with high fidelity. RNA directs protein synthesis at lowest cost. For this reason RNA was selected for containing uracil while DNA was selected for containing thymine. See Rosenberg, 1985, chapter 3, for details.

9. Kitcher, 1993, p. 172.

5

What Happens to Genetics When Holism
Runs Amok?

More than a generation of active analysis of the structure and foundations of genetic and evolutionary theory has produced more unsettled questions than it has generally accepted conclusions. Even such apparently straightforward matters like the definition of fitness and the cognitive status of the principle of the survival of the fittest remain vexed. More issues such as the number of levels of irreducible organization, the units of selection, are becoming more complex as research proceeds. As for the relations between molecular genetics, classical genetics and the rest of biology, the most widespread view attempts to combine a thoroughgoing physicalism with a resolute antireductionism. The controversy about whether there are biological laws remains with us, as do basic questions about the relation of models in population biology to one another and to some actual or possible theory.

All of these debates presuppose a series of distinctions fundamental to common sense, biology, and its philosophy: distinctions such as those between innate and acquired, organism and environment, genetic information sources and environmental informational channels, gene versus individual, individual versus group, development versus inheritance, replicator versus interactor. If these distinctions do not divide nature at the joints, then it is unlikely that debates carried on in their terms will illuminate much more than our conceptions about nature, as opposed to biological processes themselves.

A recent paper by Griffiths and Grey (1994) in effect sets out to undercut all these distinctions, and to thereby undercut all of the biology based upon them. In particular, their arguments threaten the nomological possibility of a subdiscipline of genetics – both as a separate study of heredity and a separate study of development. In this chapter, I expound their argument, strengthen it and defend it against attacks seeking to preserve the possibility of a nomological subdiscipline in hereditary and developmental genetics, and to explore the significance of their conclusion for the cognitive status of genetics.

1   UNDERCUTTING THE NATURAL KINDS
OF CONVENTIONAL THEORY

The one subdiscipline in biology which appears closest both in substance and nomological prospects to the character of physical science is genetics. Genes have characteristics that make them promising candidates for natural kinds. As Dawkins points out, they have longevity, fecundity and fidelity (1976, p. 37). But longevity and fecundity are certainly not traits limited to genes. Some environmental conditions are longer-lived than any nucleic acid or other vehicle for heredity. For example, ammonia, or methane molecules, have been around longer than genes, and lots of things have greater fecundity – even some molecules synthesized prior in the order of time to DNA may catalyze or otherwise foster the production of more tokens of their own type. But fidelity, the property of faithfulness in information storage and transmission, has long appeared distinctive of genes. We have uncovered some important regularities about genes as the bearers of information that guide development of organisms, and that are transmitted across generations. Mendel's "laws" of independent assortment, and their successors, when partially corrected for omissions, exceptions, and counterexamples, provide an account of hereditary transmission, while the so-called central dogma of one-way transmission of information from DNA-to-RNA-to-Proteins and the genetic code seemed well established across the entire range of biological development.

But whatever casual role the gene has with respect to heredity or development is duplicated by that of other non-genetic factors in heredity and development. It is obvious that the genetic material is at most causally necessary for the effects often attributed exclusively to it; other conditions must obtain for its informational role to come into play. The relationship between genotype and phenotype is highly indirect, and involves not just links in a chain from DNA to cellular assemblies, but a vast web or network of polymerases, promoters, suppressors, ribosomes and ribosomal RNAs, transfer RNAs, messenger RNAs, amino acids, etc. Within this network, the claim that the role of the gene is causally unique cannot be sustained. In particular, for any generalization about the causal role of the gene, there are any number of other generalizations about the causal roles of non-genetic items that are equally robust or rather equally lacking in robustness.[1]

But maybe what distinguished genes from other items is not their causal role but their informational one? Are genes unique information carriers? Griffiths and Grey distinguish two senses of information, and claim that neither of them sustains a special informational role for the genetic material. The first is the notion of information as it figures in the mathematical theory of information due to Shannon and Weaver. This theory distinguishes between signals and channels, and probabilizes the accuracy of information transmitted through channels subject to disturbance and distortion. Perfect information in the Shannon-Weaver theory is simply co-variation of signals sent and received.

It is natural to treat a gene in generation 1 as expressing a signal which is transmitted by reproduction to a gene in generation 2, or is transmitted by development from the generation 1 gene to a phenotype in generation 1. In this latter case, cellular meiosis and embryogenesis are channels which subject signals to varying distortions that can be measured in units given by information theory: But, Griffiths and Grey argue, the trouble is that *mutatis mutandis* genes are channels, and meiosis or embryogenesis are signals. Consider: the sequence of steps in meiosis at generation 1 is a signal transmitted by the genes to the steps of meiosis in generation 2. The presence of certain amino acids in the embryo transmits a signal through the channel of the gene. This signal is received as the synthesis of a particular protein at the stage of embryological development where a particular gene is turned on. Just as many features of a gene's environment figure as standing background conditions which the gene requires in order to transmit signals, similarly, the gene is part of the standard background conditions in which many other features may take it in turn to be the informational source and target. Applying the mathematical theory of information in fact shows that there is no clear difference in kind between any two elements of the hereditary or developmental systems as regards Shannon-Weaver information bearing.[2] Of course the non-existence of a difference in kind leaves open the possibility that there may be a substantial difference of degree. This is a possibility that we shall address at length later. However, since the rhetoric of genetics is usually expressed in terms of a difference in kind, it is fairest to explore this possibility first.

The most obvious alternative basis for a difference in kind is a semantic or intentional sense of information with which we can endow the genetic code, and the codons of nucleic acid bases that are shared in common and produce the same gene-products from generation to generation across vast ranges of species, and indeed phyla and kingdoms.[3] Perhaps, this informational role gives the genes a character distinct enough to ground distinctions that the laws of a separate subdiscipline like genetics needs.

In their reply to this ploy Griffiths and Grey go seriously wrong. Getting matters right however simply strengthens their ultimate conclusion. Griffiths and Grey simply assert that the semantic or intentional sense of information is also appropriately attributed to messages carried by non-genetically encoded non-genes. Their reply concedes that genes are information carriers. In fact, though the semantic or intentional sense of information is often employed to describe the role of genes, and their products, there is no intentionality to be found here and the use of intentional vocabulary is always metaphorical. The proof of this is that the messages these molecules contain allow full substitutivity *salva veritate* of co-referring terms and co-extensive predicates. Real intentionality precludes this sort of extensionality. Of course the fact that statements about macromolecules are extensional does not deprive the metaphor of information of its value in theorizing, but it means that genes do not literally carry information. Accordingly we cannot erect a distinction between the genes and the conditions in which they operate on the basis of their role as privileged carriers of information.

Sterelny, Smith and Dickison (1996) have argued, against Griffiths and Grey, that genes can be distinguished from other biologically significant systems on the strength of their informational roles. They too wrongly assume without argument that the attribution of intentional properties to genes is literal and not metaphorical. They write:

> We think the genome does *represent* developmental outcomes. For representation depends not on correlation but function. The plans of a building are not the primary cause of a building or its features. A plan may correlate better with graft, waste and overspending than with the actual traits of a building. Despite failures of correlation, and despite correlation without representation, plans represent buildings because that is their function. Some elements of the developmental matrix – the replicators [genes] represent phenotypes in virtue of their function. (p. 387)

Sterelny et al. argue that genes differ from other information carriers because unlike others, they were *shaped by selection* to carry information. Their function is high-fidelity information storage and transmission both in development and descent. They agree that both genes and environmental factors correlate equally well with developmental outcomes within and across generations.

> But there is an important asymmetry. . . . Consider a facultatively desert-adapted shrub; a shrub whose leaf structure and shape reduces water loss if grown in arid environments. Both aridity and the shrub's genome are necessary for that shrub's adaptive response to the environment. But that genome only exists

because of the causal path (in the genetic environment) from arid conditions to desert-adapted shrub. One element of the developmental matrix exists only because of its role in the production of the plant lineage phenotype. That is why it has the function of producing that phenotype, and that is why it represents that phenotype. (Sterelny et al., 1996, p. 388)

If only matters were so simple. The inference from function to representation is wishful thinking in which naturalistic philosophers of psychology have long engaged. To see the problem, go back to Sterelny et al.'s analogy between the blueprint plans of a building and the nucleic acid "plans" of a phenotype. The building plans represent because they are instances of "derived," not original intentionality. Plans represent because of the employment of conventions of intentional creatures like us who interpret and execute the plans. Excluding the hypothesis of divine design, there are no intentional agents whose interpretation of the nucleic acid sequences makes them representations of a phenotype. If we could show exactly how our literal intentionality was itself the outcome of natural selection, then we could hold that representation is a matter of function, just because natural selection is what creates functions. And then we could infer from function to representation both among blueprints and genomes. Alas, we know only too well that no one has yet successfully articulated this argument. Doing so would be nothing less than reducing intentional states to extensional ones. For representation is intentional and bio-function is extensional.

So, Sterelny et al.'s story about the arid environment and the shrub's genome overreaches itself. We can agree that the genome exists today because it has had the function of producing the arid-environment–adapted phenotypic leaf structure. That's just an example of evolutionary explanation. We cannot go on from there to add that it also represents or carries information in any semantic sense about that phenotype. But that's what we need to do if we are to "privilege" the genome (or an extended class of replicators) as literal information carrier. The reason is that other biological systems exist because of their functions, including many items just as causally implicated in the production of the plant lineage's phenotype as the plant's genome is. Indeed, there will be other systems selected for their causal role in the production of phenotypes. What is more, there is no single set of genetic and/or nongenetic components selected for and only for the production of the shrub's particular adaptive response to the environment. Selection for function at this level is far too blind to structure to make any such unique package. And in any case, even if some single type of genomic structure were selected for and thus had the function of producing the phenotype in question, it could no more be said to represent or carry information about the phenotype than we can say that rain clouds literally *mean* rain.

It seems to us that Griffiths and Grey rehearse the first stage in this argument in their reply to Sterelny et al. in Griffiths and Grey (1997). Here (p. 483) they emphasize that the extended replicator theory implied by the function criterion for information actually incorporates the entire developmental system as a replicator, and is therefore not any less holistic than their own proposal. The crucial step in this argument is to highlight that any trait may have multiple functions, including some which are developmental. They cite Sterelny et al.'s own example of the hand, which has not merely an economic function as the extended replicator theorists would like to claim, but also a developmental function in child rearing. However, on their account it seems like mere coincidence that the identification of replicators by means of function does not actually pick out anything but the entire developmental system. For us, however, this is because function simply cannot provide any basis for intentional information, hence it is not surprising that it fails to identify an informative class of features.[4]

It turns out then that there are even stronger arguments than Griffiths and Grey realize for their denial of a special role – informational or otherwise – for the gene. But if the gene's role in heredity and development is not causally or informationally distinctive enough to constitute a natural kind, what hope is there for any other element in biology's domain to constitute the ontological basis of a nomological subdiscipline. In the biologist's language, what hope is there that we can uncover simple, precise, and predictively useful generalizations about anything, if we can't discover such generalizations in genetics? If we seek such laws, what is the appropriate unit of inquiry?

Griffiths and Grey's answer is that the only defensible unit of inquiry in biology is the developmental system, or rather one "life-cycle" iteration of the ever-repeating developmental process. "Developmental system theory provides an alternative explanation of transgenerational stability of form. . . . The processes which replicate themselves are those which find appropriately structured resources in each generation" (1994, p. 283). Biological observation reveals cycles of a variety of packages at varying sizes that show remarkable qualitative similarity to the predecessors that gave rise to them. These packages include all of the components individually causally necessary and jointly causally sufficient for the persistence of the package – including features conventionally identified as parts of the environment and not parts of a developing or developed organism. The parts of a developmental system will be just those whose presence in each generation is responsible for the characteristics that are stably replicated in the lineage. For example, we might consider two influences on a newborn bird. The interaction between the newborn bird and the song of its own species, which occurs in each generation and helps explain how the characteristic song is produced, is part of the newborn bird's developmental process. (p. 286)

In short, anything that interacts with anything in a way that has an adaptational outcome for a lineage of life-cycles will turn out to be a part of some developmental process. ("The interactions that produce outcomes with evolutionary explanations are parts of the developmental system" [p. 286].) There is a complication here that needs to be noticed, in order not to immediately attribute holist excesses to the life cycle. Certainly Griffiths and Grey (1997) claim that this is a misunderstanding on the part of Sterelny et al.:

> The general claim that "not every reliably recurring factor is a replicator" is misleading. It implies that DST says that every reliably recurring factor *is* a part of the developmental system. DST does not claim this. It does claim that every event which is part of the evolved life-cycle is an outcome of development. That is surely a truism. But DST does not claim that everything that regularly occurs in the life-cycle is an input to the development. The inputs to development are the developmental resources. These are the things which must be produced by previous generations in order for the next generation to develop. (p. 484)

However, it is unclear to us whether this distinction between events that are outcomes of development and things that are inputs to development is that significant.[5] For it seems to serve only to rule out as developmental inputs only two categories of event that are part of the life-cycle. The first are those that are individual idiosyncrasies that are not to be explained by evolution, and the other is the persistent resources. For instance, sunlight, and gravity would seem to be parts of the developmental processes because they are persistent and have a recurring role in the adaptational outcomes that are repeated from life cycle to life cycle of many developmental systems (Griffiths and Grey, 1994, pp. 290, 291). "Resources are parts of developmental systems because of *generalizations* about their role in producing evolved outcomes" (1994, p. 297, emphasis added). However, Griffiths and Grey seek to rule them out as elements of the developmental system because they have "no causal dependence on the life-cycles they help generate." Although the persistent resources are excluded from the developmental system Griffiths and Grey are explicit that the relationship forged between the persistent resource and the organism does belong to the developmental system. We will explore the extent to which this move is effective in resisting excessive holism later.

Griffiths and Grey tell us that transgenerational stability of form is not the result of the action of the genes but the result of all the resources that are required for development: "It makes no more and no less sense to say that the [non-genetic] resources 'read off' what is 'written' in the genes than that the genes read off what is written in the other resources. . . . The theory aims to provide an explanation of generational stability of form which does not attribute it to the transmission of a blueprint or program in the genome – a pseudo explanation that inhibits work on the real mechanism of development" (1994, pp. 284, 286). It is not a developmental "system" as such which is the focus of developmental systems theory, but rather a developmental process which is segmented by iteration into "life cycles" of an evolving lineage:

> The developmental process is a series of events which initiates a new cycle of itself. We conceive of an evolving lineage as a series of cycles of a developmental process, where tokens of the cycle are connected by the fact that one cycle is initiated as a causal consequence of one or more previous cycles, and where small changes are introduced into the characteristic cycle as ancestral cycles initiate descendant cycles. The events which make up the developmental process are developmental interactions – events in which something causally impinges on the current state of the organism in such a way as to assist production of evolved developmental outcomes. The things that interact with the organism in development are developmental resources. Some of the resources are products of earlier cycles of the process, others exist independently of the process." (1994, p. 291)

The forces of natural selection themselves assure us that the biological will arrange itself into the periodicities developmental systems theory requires. Ironically, the argument for this claim is due to Griffiths and Grey's chief target, the exponent of the selfish gene, Richard Dawkins: Evolution requires repeated life cycles during which structures are repeatedly assembled and thereby exposed to variations in assembly that can be conveyed to the next life cycle. By contrast a system that merely grows ever larger in size as opposed to reproducing itself in the same size but in larger numbers, provides no scope for reorganizations reflecting improvements that can spread until every member of a given generation of the lineage reflects the improvements.[6]

Having already annulled the gene/environment distinction, Griffiths and Grey draw some further conclusions from the developmental systems approach for evolutionary theory as a whole: The most fundamental change is the denial of a theoretically grounded distinction between organism and environment. On a traditional view, individual organisms with variant phenotypes are exposed to an independently existing environment, whose fixed characteristics are only rarely shaped by the presence of these variants. In developmental systems theory, there is still variation, replication and competition. But the life cycles which vary through generations, persist and adapt, do not do so because of a greater correspondence to some preexisting environmental feature. For the life cycle includes both organism and environmental features. Nature selects for those life cycles which bear features that interact so as to provide their life cycles a greater capacity to replicate themselves than life cycles that lacks features which adaptively interact. "Organism and environment are both evolving as an effect of the evolution of differentially self-replicating life cycles. Life cycles still have fitness values, but these are interpreted, not as a measure of correspondence between the organism and its environment but as measures of the self-replicating power of the system. Fitness is no longer a matter of "fittedness" to an independent environment" (1994, pp. 299–301). We return to this conception of fitness and of evolutionary theory in section 3 below. Meanwhile, we endorse their claim that no principled distinction in biological processes underwrites a separate subdiscipline of genetics.

2 CAN WE PREVENT HOLISM FROM RUNNING AMOK?

Sterelny et al. fear that in Griffiths and Grey's hands, developmental systems theory is threatened with holism run amok. Though they concur in some of Griffiths and Grey's criticism of distinctions laid down in traditional evolutionary biology, they think that at least one of them – the replicator/interactor distinction – can be preserved, and that this distinction is enough to make a sub-science of genetics possible. Sterelny et al.'s arguments begin with their claim that the representational character of genes derives from their functions. As we have seen, these arguments are defective. As we shall now show, if there is a replicator/interactor distinction to be made in biology, it is both relational and ubiquitous and no obstacle to Griffiths and Grey's holism run amok.

Sterelny et al. note that

> a *prima facie* problem for this conception is its apparent commitment to holism. Everything is causally connected to everything else. Even so, we can understand something without understanding everything. So, if developmental systems include everything causally relevant to development, they are too ill-defined to be a coherent active unit; they are too diffuse to be the objects of selection.... Some causal influences are going to be part of the developmental system that

made us, and others are not, on pain of all developmental systems reducing to one. That would be holism run amok. (1996, p. 382)

As noted above, Sterelny et al.'s main objection to the seamless web of causally equal forces spun by developmental systems theory is based on the claim that genes do after all have a special informational role in development. This special role is owing to the fact that they are the only factors in developmental systems that have been selected for informational storage and transmission in development and heredity.

According to Sterelny et al. if there is a class of tokens selected for the function of carrying information with high fidelity and directing the development of other entities through the "publication" of this information, then there is a class of "replicators" after all. And if there are replicators, then there must be interactors – tokens whose behaviors have among their effects the differential perpetuation of replicators.[7] And of course the obvious candidates for interactors are the organisms that the genes are selected for carrying information about. In which case, there may be enough distinctiveness among the factors developmental systems theory hopes to assimilate for separate subdisciplines to exist and discover regularities. So Sterelny et al. seem to conclude.

There are however many difficulties with this argument. To begin with of course, it is regrettably false that genes have been selected for their bio-functions in heredity and development. Ninety percent of the genome in most eukaryotic species is called "junk." It is so called because though it replicates itself very nicely, and even duplicates, transposes and does other things to assure its persistence and multiplication in the genome, it has no clear role in heredity and development, or at least no role known to molecular biology. It is "selfish" DNA, which simply free-rides on the workhorse genes that do code for gene products that make adaptative phenotypes. "Selfish" DNA has been selected for its own replicatory powers, and for no further bio-function – informational or non-informational. Of course we could simply deny that selfish DNA sequences are genes, but this fiat will simply read them out of developmental systems and so leave them out of biology altogether. It is a ploy with the advantages of theft over honest toil.

More important, as we have seen above, the inference from bio-function to information is a vexed one. Sterelny et al. cannot help themselves to the conclusion that the genes are selected for their informational role in heredity and development unless they can upgrade extensional functions into intentional representations.

Let's consider whether Sterelny et al. really need so strong a claim in order to make out a special role for the gene, and thereby to ground a series of distinctions that developmental systems theory purports to efface. Might it be enough to claim merely that the genes are uniquely selected for some non-intentional functions; not information storage and transmission, but the storage and transmission of instructions for the synthesis of polypeptides where these instructions allow substitution *salva veritate* in their terms and predicates? The trouble with this idea is that too many things will count as replicators of a given developmental system, or else only one thing will count as a replicator – the developmental system itself. This is the message of Griffiths and Grey's argument. Any factor that regularly recurs in single life-cycle tokens of the developmental process's lineage and that has adaptational benefits for the lineage's members will have a function in heredity and development – this will include genetic material, but also factors like maternal RNA and proteins, environmentally altered organelles in asexual species, or the shape of a bird's nest (Bateson, 1978). Indeed, Sterelny et al. provide an extended example of a non-genetic replicator: the penguin nesting burrow.

> [V]ariation in a burrow can cause a variation in a burrower: a particular penguin chick may be healthy and safe because its burrow has one site rather than another, even though no burrow features make penguin flesh.
> [B]urrows are parts of a copying and interaction cycle. They exist in the forms they do because of their role in this cycle. . . .
> Burrows bear information about burrows and the next burrow generation. For burrows interact with their guests in a way that results in mutual changes. They and their guests co-evolve. Chance changes in burrow copies can proliferate. A chance favorable burrow copying at a new and superior site – for example, one less liable to flood – may result in a bushy lineage of similarly changed burrows. If nests [or burrows] are replicators, they are clearly active replicators. Their properties of insulation, durability, protection and costs of construction quite obviously influence their probability of being reproduced. They form lineages. A nest plays a role in the construction and protection of builders who disperse to produce a new and relevantly similar nest. There is a flow of information linking nest generations through the builders. Nests and burrows are adapted for the growth of burrow builders and nest-makers. Those interactors carry the information through which the nest is replicated. (Sterelny et al., 1996, pp. 398–399)

Everything here is right except the claim about burrows bearing information. They do not, but they are selected for their adaptive consequences for their "occupants" and thus ultimately their descendant burrows or nests. If this is enough, then it will turn out that most biological systems are both replicators and interactors for a vast number of other bio-systems. For consider, if nests and burrows are replicators, then so are all the other adaptive phenotypes of the birds which nest and burrow. For nests and burrows are nothing but what Dawkins calls the extended phenotype. And it's not just the phenotypes of a species that will be replicators. A symbiotic species will turn out to be a replicator for the species on which it so functions. For the symbiot is selected in part for what it does for the symbiot. Indeed, any two co-adapted species, populations, kin groups, etc., will be one another's replicators. And we may generalize this result: to the extent evolution leads to equilibria among large numbers of species, close to equilibria everything biological will be a repli-cator for everything else biological in the biological system. And *mutatis mutandis*, everything will therefore be an interactor for everything else in the system. Any lineage of a developmental system will have as many replicators as it has successful interactors, for they will be one and the same. And lineages in equilibria will have at least as many kinds of replicators and interactors as there are other lineages in their equilibria. This is pluralism run amok, or alter-natively, it is just a way of expressing the notion that the replicator/interactor distinction is a relational one. Or alternatively, if the replicator/interactor dis-tinction is not after all a relative one, then the only intrinsic non-relative replicators and interactors will be the largest developmental system. Whence threat of holism run amok.

3   WHAT EVER HAPPENED TO THE THEORY
OF NATURAL SELECTION?

What follows on the developmental systems' approach for the nature of evo-lutionary theory? As we noted above, Griffiths and Grey claim that "taking developmental processes, rather than genes or transitional phenotypes, to be the units of evolution requires a substantial reformulation of evolutionary theory" (1994, p. 300). This may indeed be an understatement. For a start it requires that some evolutionary stories be Lamarckian in form. This conse-quence is noted by Sterelny et al. (1996, p. 392) and accepted by them as a consequence of their extended replicator theory as well. As they note this is not an unacceptable consequence:

> Weismann and his successors have shown that one apparently possible mecha-nism of evolutionary change is not actually possible. An organism's phenotype can change and change in a way that alters its genotype: a mouse can shift its residence to a leaky nuclear reactor. But an organism's changes cannot restruc-ture its genotype so that its descendants manifest the changed trait. A mouse that acquires the ability to exploit a new food source cannot transmit that ability

108

to its descendants via changes in its genome. The discovery of this constraint on inheritance is of great significance. But the discovery is the discovery of a constraint on a specific mechanism of inheritance, not a constraint on any mechanism of inheritance. Moreover, Weismann did not show that the only inheritance mechanism is genetic; few deny that social learning is an inheritance mechanism. (1996, p. 392)

Genotypes might not be capable of the passing on of acquired characteristics but the numerous other resources involved in the developmental system, such as behaviors passed on through social learning, or locales of habitation suffer no such restriction. However, while this is a dramatic expansion of the possible evolutionary explanations, another point suggests that the favored method of evolutionary explanation, natural selection, may be dramatically affected. For as we see it, the developmental systems approach abridges yet another distinction, without which the central insight of evolutionary theory, adaptation, is unstateable.

Griffiths and Grey call into question "the separation of organism and environments." But by doing so developmental systems theory leaves no conceptual space for the possibility of increasing local adaptation of a lineage of organisms to a distinct and separate environment which tests variants and selects the fittest of them. Of course, developmental systems theory might allow for distinctions between components of life cycles – say, the organismic one and the environmental one – and permit us to trace the presumably faster changes in the former, showing how they increasingly adapt themselves to the latter, thus enhancing the "fitness" of the whole package. But what is the difference between a distinction drawn within developmental systems theory for this purpose and the traditional claim that evolutionary biology is about the evolution of lineages of organisms in distinct environments? Griffiths and Grey's innovation cannot be a matter of the pluralism now widely advanced among philosophers and biologists – the thesis that there are many levels of organization – many different life cycles – genes, individuals, kin-groups, populations, species, ecosystems – nested, overlapping, competing. For this thesis requires all the distinctions that Griffiths and Grey seek to extirpate.

One might read Griffiths and Grey's attempt to exclude persistent resources from the life cycle as trying to restore some form of organism–environment distinction. If the persistent resources genuinely can be excluded from the life cycle then they would seem to be able to constitute something like an environment. We have already indicated that we have some more metaphysical worries about the relationship between the life cycle as a repeated sequence of events and the resources as persisting things. Beyond these worries, however, there seem to be good reasons for doubting that the persistent resources can serve as anything like the environment that is involved in evolutionary explanations. The main basis for this is that the best candidates for persistent resources are things such as sunlight or gravity that are genuinely independent of the action of any life cycle (although it is of course "up to the life cycle" whether it chooses to take advantage of the resources). Griffiths and Grey often write as if the persistent resources were a much larger class than this. For instance, they cite as an example the shells that hermit crabs "borrow" as an example of a persistent resource. This seems mistaken to us. For while it may be true that no part of the life cycle is causally responsible for the creation of this resource, it is responsible for its consumption. As Griffiths and Grey note, species of hermit crab that use only the shells of extinct species are condemning themselves to eventual extinction. If the hermit crab life cycle can be responsible for the consumption of the persistent resource in this way, then it seems unable to act as an environment for an evolutionary explanation. The more limited class of resources that are genuinely persisting, in being immune from biological consumption, such as gravity and the energy of the sun, might be capable of serving as an environment. But these also have a severe weakness as regards conventional evolutionary explanation: they are available for all the species of the Earth. The traditional explanation of the specialization of the creature to fit a particular environment is lost, if all creatures inhabit a similar environment.

Thus we think that Griffiths and Grey capture the true form of evolutionary theory under developmental systems when they write that:

> In the developmental systems representation, the variants differ in their capacity to replicate themselves. One variant does better than another, not because of a correspondence between it and some preexisting environmental feature, but because the life cycle that includes interaction with that feature has a greater capacity to replicate itself than the life cycle that lacks that interaction.... Organism and environment are both evolving as an effect of the evolution of differentially self-replicating life cycles. Life cycles still have fitness values, but these are interpreted, not as a measure of correspondence between the organism and its environment, but as measures of the self-replicating power of the system. Fitness is no longer a matter of "fittedness" to an independent environment. (1994, pp. 300–301)

To see whether this conception of fitness is viable, ask how the differential fitness of a life cycle is to be explained. Fitness is a comparative concept in evolutionary biology, not an absolute one. One system is more fit than another, and this difference is measured in differences in their expected reproductive rates. Now life-cycle tokens can differ in this way, provided that the fitter among them is the one that leaves more tokens. The explanation developmental systems theory gives for this difference will be that the components of the fitter life cycle token interacted with one another in such a way as to produce more tokens than the less fit life-cycle token. But what made the more reproductively successful interaction fitter? It cannot be that one token's feature interactions were more well suited to the environment in which it obtained than the other token's feature interactions were. For that would be to introduce the rejected "correspondence" between independent environment and life cycle. Is there another causal foundation that will explain differences in fitness among life cycles?

Short of an explanation of why some feature/interactions are fitter than others, the explanatory power of fitness differences will be reduced to zero. For without appeal to a distinct environment we are explaining reproductive differences in terms of fitness-differences; but these fitness-differences go unexplained as basic differences between life cycles without underlying causal sources. Griffiths and Grey describe fitness as a measure of the self-replicating power of a system (1994, p. 301). Unless this power is like Voltaire's dormitive virtue it requires an occurrent base in non-dispositional properties of life cycles. And of course fitness is always relational and comparative. Organisms, genes, life cycles, whatever, are always more or less fit than others; never simply fit *sans phrase*. And they are fitter than others with respect to constraints under which they exist. Everything we know about self-replication suggests that systems which engage in it do so by tapping sources of energy and order beyond themselves. Accordingly developmental systems theory cannot escape the ultimate demand that fitness be explained "as a matter of 'fittedness' to an independent environment" (1994, p. 301). To deny this, as developmental systems theory seems to require, is to deprive evolutionary theory of its explanatory content. This would be more than a "substantial reformulation of evolutionary theory."

Suppose the fitness of a token is not a matter of "fittedness" to something distinct from the token. Suppose it is a monadic property supervenient on properties of the token and its parts, but not supervenient on relations between the token's parts and something distinct from the token. Then differences in fitness between tokens will not be detectable by locating differences in "engineering design" – differences in the way systems solve survival and reproductive problems posed by an external environment. They will only be detectable in the phenomenon that fitness-differences are meant to explain: longevity and differential reproduction. We will be faced with the hoary old problem of triviality and tautology. (There are of course philosophers who hold that the claim that the fittest survive is a tautology, but they find the explanatory power of the theory in its claims about adaptedness of tokens to distinct environments – something developmental systems theory cannot accept.)

The upshot is that so far as the structure of the theory of natural selection is concerned, the developmental systems approach either leaves things pretty much as they were, or else requires a completely new account of evolution. One thing it doesn't do is simply "require a substantial reformulation" of the theory. On the former alternative, which preserves the core conception of natural selection as the filtering of variants by an independently existing environment, developmental systems theory is simply the claim that the unit of selection is the life cycle of a much larger package than most evolutionary biologists recognize. On the latter alternative, which replaces evolutionary theory altogether, the result is as yet impossible to describe. We don't have a replacement for this theory.

4   COPING WITH HOLISM RUN AMOK

We believe that Griffiths and Grey are correct in their claim that most of the fundamental distinctions in evolutionary biology are not natural kinds, do not divide nature at the joints. Moreover, we believe that though developmental systems theory is holism run amok, this sort of holism really does characterize the biosphere. What are the implications for the actual practice of biology?

Begin with molecular genetics. Griffiths and Grey are correct. The gene is not a natural kind, and the notion reflects an artificial distinction among a range of informationally and causally equal factors in heredity and development. But some factors are more equal than others. Genes may not differ in kind from other causally relevant factors in heredity and development, but they differ in at least one important degree. The role of genes much more closely approaches the role of literally intentional information bearers than does anything else among the class of factors causally necessary for hereditary and developmental processes. Though ascribing intentional states to genes remains a metaphor, it is a live one and a much livelier one than the same metaphor employed elsewhere. By livelier we mean that employing the metaphor of literal, intentional information bearing, biologists have been able to develop useful tools of very great power. Population and molecular genetics is certainly the area of biology richest in models and generalizations.

For a long time molecular biologists believed that they had hit upon firm nomological generalizations about genes. Had they done so, the claims of developmental systems theory would have been refuted by science, not philosophy. Among these putative laws there was the claim that in our biosphere at any rate, all hereditary "information" is exceptionlessly carried by DNA (the scare quotes remind us that we have no right to this claim as a literal description); there is a single genetic code for protein. Or again, as the ironically so-called central dogma has it, the direction of information flow is one way: DNA is transcribed into RNA, and RNA is transcribed into proteins. Third, all enzymes are catalytic proteins (amino-acids). These three claims, initially held to be exceptionless, have turned out to be almost true, but not quite. First it was discovered that there are the RNA viruses which carry their genetic "blueprints" in RNA; then it was discovered that RNA can catalyze its own splicing, and so counts as an enzyme; now it is becoming clear that some small proteins – the prions responsible for Mad Cow Disease – can transmit hereditary information from generation to generation without recourse to nucleic acid at all.

Developmental systems theory helps us see why even at the basement level of biology selection makes everything interact with everything else, spoiling neat generalizations, and undercutting the possibility of nomologically natural kinds. Because selection is constantly moving through the space of continually expanding variations, filtering out maladaptations, any adaptational strategy that one can imagine seems to be realized some where and some when in creation. No matter how bizarre a strategy for survival seems, no matter how intricate a series of behaviors it requires, that strategy seems exemplified at least once. With the broadest constraints, no matter how inhospitable an environment seems – no matter how cold or hot, arid or dark, toxic or radioactive, there is a niche to be filled; and for every initially apparently robust generalization about the adaptation of particular life cycles of particular lineages, there turn out to be serious counterexamples, exceptions, and alternative ways that nature has found to skin her cats. There has been world enough and time for creatures with our powers to emerge. That time has also been sufficient for an overwhelmingly large number of different niches to be filled, for every adaptive strategy to spawn a counterstrategy, and for every generalization about any of these strategies to be riddled with exceptions.

Now if this is correct, then the sort of causal independence that autonomous levels or units or tokens require almost never obtains. If anything, nature honors a principle of the plenum. Over the long run all biological possibilities are canvassed, and everything biological is causally implicated in the shaping of everything else biological. Thus, above the smallest macromolecule any generalization will have endless exceptions, reflecting all these possibilities. When we reach the level of framing mathematical models – the stock in trade of population genetics – it will be no wonder that there is no general underlying theory that unifies and explains when these models work and when they don't. We do know when they work and when they don't, but only ex post. They are confirmed in some data sets and disconfirmed by others, and we can even identify the particular causal forces that interfere with their empirical adequacy. But unlike physical theory, we lack a general theory which unifies these models. And now we see why we lack such a theory. If everything interacts with everything, there will be nothing more than empirical regularities; there will be no laws, no natural kinds into which biological phenomena are subdivided. The result is not pluralism, or indeterminism. Pluralism requires the existence of many "units" or "levels" at which autonomous generalizations obtain; indeterminism is the absence of causality. In biology the problem is too much causality. What we are really faced with is holism run amok. And what Griffiths and Grey have supplied is another explanation for why it obtains.

Because developmental processes will embed one another every time embedding is a useful strategy for survival, Griffiths and Grey credit developmental systems theory with a pluralism widely claimed nowadays: "There may be several levels of life cycles, accounting for different features of evolved systems. We are suspicious of the term 'level' here, however, since investigation has not yet proceeded far enough to determine to what extent processes at one 'level' can be considered independent of those at other 'levels'" (1994, p. 295). But the whole thrust of developmental systems theory is that instances of every biological kind are potentially, and almost always actually, linked up to instances of every other type closely enough to make a difference for survival strategies and adaptive consequences. Even where almost every token of a given type may be independent of tokens of every other type, there will be one or more lineages of life cycles more or less directly linked to enough others to make them parts of one another's life cycles or developmental processes. This is holism run amok. Evolutionary outcomes appear to sustain a sort of principle of the plenum: Every adaptive strategy has a consequence for the environment of so many other adaptive strategies, that all interact over evolutionary time scales. And this much interaction makes timeless truths in terrestrial biology impossible to come by.

Development systems theory really leaves evolutionary biology pretty much as it found it, employing all the distinctions which Griffiths and Grey show to be unwarranted by biological processes themselves. All developmental systems theory does is show that biological theory is thoroughly heuristic, an instrument useful to agents of our cognitive abilities and interests, a theory whose predictive weakness and explanatory imprecision the developmental approach illuminates, but not one the developmental approach can replace, at least not for creatures with our cognitive limits and practical interests.

The intentional stance is apt in molecular biology because here extensional nature comes as close to intentionality as it can possibly come. Note that the functions of the human brain will be subserved by the same molecular and atomic forces that subserve heredity and development. But for all that attributing an informational role to genes is a stance, a useful instrument for creatures like us, who think themselves to have intentional states, and exploit this belief in our theorizing. When we attempt so to organize the behavior of nucleic acids and the proteins they (and their environments together) create into laws and theories we inevitably fail. This is because under intentional descriptions, and indeed under functional descriptions, these putative generalizations will become weaker and more exception-ridden as time and selection for function goes on. And these generalizations will not be replaced by more fundamental and less exception-ridden ones that might explain them and their exceptions. The reason is that these generalizations – from Mendel's laws down to the central dogma and the regularities of the genetic code – don't divide nature at its joints, but reflect a metaphor not yet superseded in the advance of scientific theorizing. If we are right about human cognitive limits, it is a metaphor we may never relinquish.

Other metaphors in science have been relinquished when the theories with which they have been associated were superseded by predictively and explanatorily more powerful ones. At these points we recognized that there were no laws about heat as a fluid, or impetus, or matter as indestructible substance, and that these metaphors were now well and truly dead – that is, no longer sources of scientific advance.

But given our cognitive and computational limitations, and our own interests as living creatures selected for survival and reproduction, not for generality in theorizing, the heuristic devices of biology – the distinctions Griffiths and Grey have shown us to be biologically groundless – will continue to be useful. And this means that despite the truth of holism run amok, the distinctions between innate and acquired, organism and environment, genetic information sources and environmental informational channels, gene versus individual, individual versus group, development versus inheritance, replicator versus interactor, will remain with us, as will the empirical regularities and abstract models couched in their terms – regularities and models which will never give rise to anything recognizably nomological by the lights of physical science, but which are indispensable for all that.

NOTES

Revised and reprinted from "La génétique et la holisme débridé," *Revue Internationale de Philosophie*, 54 (2000): 44–61. Coauthored with Andrew J. H. Clark.

1. Sterelny, Smith and Dickison (1996, p. 381) offer an extended gallery of possible causal distinctions, but none of these appear conspicuously more robust.
2. This conclusion is (partially) supported by Deutsch (1996, p. 173). Although an enthusiastic proponent of the theory of replicators and interactors, he is forced by the case of junk DNA to admit that the distinction is "strictly speaking a matter of degree." However, he later invokes a supposed resemblance of genetic information bearing to virtual reality rendering and the stability of such structures across possible universes of the quantum mechanical multiverse in order to reestablish the distinction. We do not consider such "high tech" solutions to the problem, although those that seek to resist our conclusions should be aware of them.
3. There is a halfway house between the two forms explored in Sterelny and Kitcher (1988), which unpacks the informational content in terms of the genes' role in the normal total environment (subtly different to the Shannon-Weaver model which involves holding the other factors constant). Griffiths and Grey (1994) argue against this on the basis that for most genomes (their example is acorns) the normal state of affairs is a failure to develop, hence acorn genomes correlate better with rotting than anything else. The force of this criticism is conceded by Sterelny et al. (1996) who "doubt that there is a quick fix for this problem."
4. Indeed it is deeply unclear to us that it is even desirable for something like the derived intentionality of a blueprint to be a property of replicators. For it is a very significant fact that, examined closely enough, the biological universe, with all its redundancies and ad hoc modifications, does not look much like a designed system; this is the origin of all the standard rhetoric about the "blind watchmaker" and the ultimate disproof of the hypothesis of design. If it were fundamental to the nature of replicators that they carry something very like derived intentionality then it seems most unlikely that the products of evolution would have the form we observe.
5. That it confuses two separate distinctions, that between input and outcome, and event and thing, is not helpful. If DST has a weak link it may well be due to a confusion between event and thing. The life cycle is after all a set of events, while the resources that it needs to generate seem to be things. However, the relationship between a thing and an event is a non-trivial philosophical issue, and the only definite weakness we are willing to attribute to Griffiths and Grey on this score is the problem of persistent resources, which we will deal with later.
6. Griffiths and Grey's invocation of Dawkins' argument may be inconsistent, for they go on to reject one of its presuppositions: the existence of an environment distinct from the lineage which evolves and increasingly adapts to the environment. See section 3 below.
7. There is an alternative approach to such arguments that seeks to reason from the existence of a distinctive class of well-designed interactors to the existence of replicators

by some form of inference to the best explanation; the existence of genes is the only way of describing the evolution of the well-designed functional structure of the organism. The DST approach has two objections to this. First it is most unclear whether the existence of such pure replicators is the best explanation for the highly evolved stable form of these interactors, or whether it simply involves wiring in all the hard stuff about the stability of form and informational content. Secondly, the persuasiveness of the life cycle conception suggests it may well not be possible to carve off one particular part of the life cycle as the structure of the organism in the way this argument requires.

6

The Biological Justification of Ethics

A Best-Case Scenario

Social and behavioral scientists – that is, students of human nature – nowadays hardly ever use the term "human nature." This reticence reflects both a becoming modesty about the aims of their disciplines and a healthy skepticism about whether there is any one thing really worthy of the label "human nature."

For some feature of humankind to be identified as accounting for our "nature," it would have to reflect some property both distinctive of our species and systematically influential enough to explain some very important aspect of our behavior. Compare: molecular structure gives the essence or the nature of water just because it explains most of its salient properties. Few students of the human sciences currently hold that there is just one or a small number of such features that can explain our actions and/or our institutions. And even among those who do, there is reluctance to label their theories as claims about "human nature."

Among anthropologists and sociologists, the label seems too universal and indiscriminant to be useful. The idea that there is a single underlying character that might explain similarities threatens the differences among people and cultures that these social scientists seek to uncover. Even economists, who have explicitly attempted to parlay rational choice theory into an account of all human behavior, do not claim that the maximization of transitive preferences is 'human nature'.

I think part of the reason that social scientists are reluctant to use "human nature" is that the term has traditionally labeled a theory with normative implications as well as descriptive ones. Anyone who propounds a theory of human nature seems committed to drawing conclusions from what the theory says is the case to what *ought* to be the case. But this is just what twentieth-century social scientists are reluctant to do. Once the lessons of David Hume [1888 (1737)] and G. E. Moore [1907] were well and truly learned among

118

social scientists, they surrendered the project (associated with the "moral sciences" since Hobbes) of deriving "ought" from "is."

The few scientists who have employed the term "human nature" do draw evaluative conclusions from their empirical theories. The best recent examples of such writers are sociobiologists like E. O. Wilson [1978], eager to extend the writ of evolutionary biology to include both the empirical study of humans and the foundations of their moral philosophy.

It is relatively easy to offer a review and philosophical critique of the excesses that are bound to creep into evolutionary biologists' attempts to transcend the traditional limits of their discipline. But more useful than still another catalog of sociobiological foibles would be a sympathetic examination of the best we might hope for from the application of evolutionary biology to traditional questions about moral philosophy. Of all the intellectual fashions of the late twentieth century, it has the best claim to provide an account of human nature in the scientist's sense of "nature," for it is undeniable that every aspect of humanity has been subjected to natural selection over blind variation literally since time immemorial. If any one thing has shaped us it is evolution, and if any piece of science is going to shed light on ethical issues, sociobiology – the application of Darwinian theory to human affairs – will. Therefore, my aim will be to identify the minimal conditions under which evolutionary biology *might* be able to tell us something about traditional issues in moral philosophy. If the rather strong assumptions evolutionary biology requires to shed light on these issues fail to obtain, then – as our best guess about human nature – biology will have no bearing on moral philosophy. This, in fact, is my strong suspicion. Nevertheless, I herewith attempt to put together the best-case scenario for the ethical significance of evolutionary biology.

1   THE POSSIBLE PROJECTS

There are several sorts of insights evolutionary biology might be supposed to offer about human nature and its relation to morality. One among them is uncontroversial and beyond the scope of moral philosophy. Like any scientific theory, evolutionary biology may well provide factual information that, together with independent normative principles, helps us make ethical decisions. It may uncover hitherto unnoticed means we can employ in meeting ethically established ends. It may even identify subsidiary goals that we need to meet in order to attain other intrinsic goals. For example, there are plain facts (about, for example, ecology, genetic diversity, and the importance to us of preserving threatened species) which biology reveals and which can be combined with moral standards into hypothetical imperatives governing human action.

More controversially, evolutionary biology may reveal constraints and limitations on human behavior that our ethical prescriptions will have to take account of. If "ought" implies "can," the contrapositive will be valid too: "can't" should imply "need not." Like other scientific theories, evolutionary biology may help fill in the list of what we (nomologically) cannot do. However, for a theory of human nature to have ramifications for moral philosophy itself, it will have to do more than any of these things.

The most impressive accomplishment for a theory of human nature would be the derivation of particular moral principles, like the categorical imperative or the principle of utility, from biological facts about human beings. Slightly less impressive would be to derive from such facts our status as moral *agents* and subjects, or to establish on the strength of our biology the *intrinsic value* of human life. A derivation of agency or intrinsic value is equivalent to deriving the generic conclusion that there is some normative principle or other governing our actions. Such a derivation would be less impressive because it would leave open the question of which moral principles about agents or objects of intrinsic value were the right ones. Still less impressive but significant in its own right would be the derivation of some important component or condition or instance of morally praiseworthy conduct – like cooperation, altruism, or other-regarding behavior – as generally obligatory. To be significantly interesting the derivation need not be deductive, but it cannot be question-begging: it cannot begin from assumptions with substantial normative content. Otherwise, it will be open to the charge that these assumptions are doing all the real work, and that the biological theory makes no distinctive contribution to the derivation.

The possibility of this project, of deriving agency and/or value (or, equivalently, deriving the existence of some moral principle or other), rests on two preconditions. The first is that we can derive "ought" from "is": that there is some purely factual, empirical, contingent, strictly biological property of organisms, which could underwrite, explain, or justify their status as agents or loci of intrinsic value. The second is that this property is *common and peculiar* to *all Homo sapiens*, so that it will count as constituting our nature.

That the first of these two preconditions for deriving morality from human nature cannot be realized seems to me to be at least as widely held a view as any other claim in moral philosophy or metaethics. Accordingly, I will not offer new arguments to supplement the observations of Hume and Moore. I recognize, however, that the more sophisticated sociobiologists are perfectly aware of these strictures on moral justification. Among sociobiologists, those who nevertheless go on to attempt to derive some normative claims from biological findings do Moore and Hume the courtesy of noting and rejecting their arguments (cf. Wilson, 1978, chapter 1; Alexander, 1979).

But even if we grant the sociobiologist's claim that the derivation of 'ought' from 'is' has not yet been totally excluded, there remains a second precondition required by the project of deriving morality from human nature. And the failure of this condition is something on which all evolutionary biologists should be in agreement.

Humans are supposed to be moral agents. This is what distinguishes us from moral subjects, like animals, and from morally neutral objects. Now, for some biological property of human beings to ground our status as the unique set of moral agents (in our biosphere at least), that property will have to be as widely distributed among human beings as the moral property it grounds, and it will have to be peculiar to humans as well. For if it is not restricted to humans, there will be other subjects with equal claim to the standing of moral agents. The trouble is that if modern evolutionary biology teaches anything, it shows that there are no such properties common and peculiar to each member of a species. If there were, taxonomy would be a much easier subject. And since there are none, what evolutionary biology in fact shows is that there is no such thing as *the* unique human nature, any more than there is beaver nature or dodo-bird nature or *E. coli* nature.

Population genetics and molecular biology have shown that, up and down the entire range of living things, there are no interesting *essential* properties – no properties which will explain a range of behavior in the way that, say, molecular structure explains most of what a chemical compound does. It is not that modern biology has yet to find such essential properties, which give the nature of a species. Rather, evolutionary and genetic theory *requires* that biological species have no such common and peculiar essential properties.

Gradual evolution by natural selection requires vast amounts of *variation* within and between species. This variation is provided by mutation, genetic drift, immigration, emigration, and most of all by genetic recombination in the sexual reproduction of offspring. The result is that there are no *essential* (suites of) phenotypes. Neither the typical nor average nor mean nor median values of the heritable phenotypes which face selection are their *natural, essential* values. They do not constitute the normal traits of members of a species, from which differences and divergences might count as deviations, disturbances, defects, or abnormalities. Of course there are biological properties common to every member of a species. For example, all *Homo sapiens* engage in respiration. But then so does every other organism we know about.

Similarly, there are some biologically based properties peculiar to individual humans – self-consciousness, speech, a certain level of intelligence, opposable thumbs, absence of body fur, etc. But these properties are plainly not distributed universally among humans, nor would the lack of any one of them be enough to deprive someone of membership in our moral community who was otherwise endowed with it. There is no human nature in the sense in which 'natures' are identified in modern science.

It might well be supposed that there is some complex combination of properties – say, self-consciousness *cum* opposable thumbs *cum* a disjunction of blood types – that is sufficient for moral agency. But the project of grounding agency in (and only in) human nature requires that this complex combination of properties be necessary for agency as well as sufficient for it, and that it be universal among *Homo sapiens*. For consider, how could a property *restricted* in its instantiation only to some members of a class provide the basis for a property *common* to all members of the class: how can we derive "All As are Cs" from "Some As are Bs" and "All Bs are Cs"? Doubtless, a philosopher can solve this problem by cooking up some gruesome gerrymandered relational property. For example, one could define property C as the property of being a member of a class some of whose members have property B. Then the derivation is trivial. But, clearly, being a moral agent is not a relational property – not at any rate, if it derives solely from the nature of the individual human. And this makes the logical problem a grave one for those who seek to derive agency from human nature.

Deriving a particular moral principle, or even the generic status of moral agency, from human nature alone – at least as evolutionary biology understands it – is not a feasible project, even if we could derive "ought" from "is."

A third potential project for the biological account of human nature is that of *explanation*: telling a plausible story about how a particular moral principle or "morality" in general, or some important precondition or component of it, emerged in the evolution of *Homo sapiens*.

The *qualification* "plausible" cannot be emphasized too strongly here. The most we can expect of any evolutionary account of chronology is plausibility: that the narrative will be consistent with evolutionary theory and with such slim data as may be available. The reason is that the problem of explaining the emergence of morality is similar to (but even more difficult than) that faced by, say, the task of explaining the disappearance of the dinosaur. There is a saying in paleontology: "The fossil record shows at most that evolution occurred elsewhere." In the case of explanations of the evolution of behavior, there are no bones, no "hard parts" left to help us choose among competing explanations. The most we can hope for is plausibility.

This raises the question of how much a merely plausible story is worth, what it is good for, and why we should want it for more than its entertainment value. The question is particularly pressing in moral philosophy and metaethics. For it is not clear that even a well-confirmed explanation for the emergence of aspects of morality from human nature has any relevance to the concerns of philosophers. It would be a genetic fallacy to infer that a particular normative conclusion was right, justified, or well grounded – or, for that matter, that it was wrong, unjustified, or groundless – from a purely causal account of its origins.

If, however, we could parlay the explanation for the emergence of aspects of morality from human nature into an argument about why it is rational to be moral, then for all its evidential weakness, the causal story would turn out to have some interest. It would address a traditional question in moral philosophy: why should I be moral? There may be some reason to think such a strategy will work. For natural selection is an optimizing force for individuals, and so is self-interest. Except under unusual conditions, natural selection cannot operate to optimize the properties of groups, as opposed to individuals. Explanations in evolutionary biology proceed by rationalizing an innovation as advantageous for an organism's survival. Egoistic justification does something quite similar. Like evolutionary explanation, it rationalizes actions as means to ends.

This, I think, is the only interesting project in moral philosophy or meta-ethics for a biological approach to human nature. In what follows, I sketch the outlines of such a project. I should note two things about my sketch. First, little that follows is original. Mostly, I have plucked insights from a bubbling cauldron of sociobiological and evolutionary theorizing. Second, I am by no means optimistic that this project of rationally justifying morality can succeed, even in part. My aim is to identify the strictures it will have to satisfy if it stands a chance of succeeding.

2 NATURAL SELECTION, BLIND VARIATION, FITNESS-MAXIMIZATION

If the theory of natural selection is right, then the overriding fact about us is that we are all approximate fitness-maximizers. Of course, this is not a special feature of people. Indeed, it is the most widely distributed property of biological interest that there is. Every organism in every reproducing species is an approximate fitness-maximizer, for natural selection selects for fitness-maximization *überhaupt*. All the phenotypes that have been selected for in the course of evolution have this in common. And if the theory is correct, then over time, given constant environments, successive generations of organisms are better approximations to fitness-maximization than their predecessors. This is what adaptation consists in.

What exactly is fitness-maximization? This is a vexed question in the philosophy of biology. For present purposes it will suffice to adopt the following definitions: x is fitter than y if, over the long run, x leaves more fertile offspring than y. Thus, an organism maximizes its fitness if it leaves the largest number of fertile offspring it can over the long run. It will be convenient if we define 'offspring' in a special way: an offspring will count as one complete set of an organism's genes. Therefore, the result of asexual reproduction is one offspring, but the result of sexual reproduction is half an offspring, since each child bears only one-half the genes of each of its parents. Note that, by this means of reckoning offspring, if a childless woman's brother has one child, the woman has a quarter of an offspring. Thus five fertile nieces and nephews make for greater fitness than one child: $\frac{5}{4} > 1$. This means that nature, in its relentless search for fitness-maximizing organisms, sometimes selects for fewer children and more offspring.

Nature selection has made us *approximate* fitness-maximizers, not perfect ones. There are several reasons for this. To begin with, nature selects for fitness-maximization only indirectly, by seeking adaptive phenotypes: among giraffes it selects for long necks, among cheetahs for great foot speed, among chameleons for mimicry, and among eagles for eyesight. But each of these is selected because it makes for the survival and the well-being of the organisms endowed with it. And survival, along with well-being, are in turn necessary conditions for reproductive success. Mere survival is not enough; an organism must be healthy enough to reproduce and ensure the survival of its offspring. But the point is that except where selection operates directly on the organs of reproduction, birth, feeding, and protection, every other piece of an organism's equipment is selected for direct effect on survival and well-being, and through them for indirect effects on fitness. This means that much of what nature selects may not look like it bears on reproduction and fitness.

In its culling of these properties that bear indirectly on fitness, natural selection puts a premium on quick and dirty solutions to the problem of fitness-maximization. It prefers these cheap, imperfect solutions to slow but sweet ones that may do the job better but take a long time to emerge. Nature recognizes Keynes's maxim that in the long run we are all dead, and it acts on this maxim before it's too late. Thus, all organisms are at best approximate, jury-rigged, only intermittent fitness-maximizers. As genetic recombination and the other sources of phenotypic novelty turn up variations, the best among them out-reproduce the others. But the best may not be very good on any absolute scale. It need only be good enough to survive and outlive the other variants among which it emerges.

Selection operates on what variation provides. It has no power to call forth solutions to problems of adaptation, only to pick and choose among those that recombination and mutation may offer. Here is a nice example (with thanks to Daniel Dennett). Fish need to be able to recognize predators. But fish do not have very sophisticated predator recognition capacities; they have not evolved the sort of cognitive capacities for discriminating other fish, let alone telling friend from foe. Yet in the presence of predators they invariably startle, turn, and flee. Of course they also respond this way to all fish, not just the predatory ones – indeed, present a fish with any bilaterally symmetrical stimulus and it will emit this flight response. The reason is that selection has resulted in the emergence of a relatively simple solution to the predator detection problem: bilateral symmetry detection. This is not a very discriminating capacity, but at least it is within the cognitive powers of a fish, and it works well enough at predator detecting. Its defects are obvious – the fish wastes energy fleeing non-predators. But in its environment the cost of this imperfection is low enough, and without it fish would not have lasted long enough to give rise to those species cognitively powerful enough to do the job of predator recognition any better.

There is a related reason why natural selection leads to the evolution only of approximate fitness-maximizers. Environments change, and organisms must survive in an environment that manifests wide extremes, and they must survive when one environment is displaced by another environment. Such conditions put a premium on being a jack of many trades instead of a master of one. An environment of great uniformity lasting over epochs of geological length provides selection with the opportunity to winnow successive variations to remarkable degrees of perfection. Consider the human eye, which is the result of a series of adaptations to a solar spectrum that has remained constant for almost the whole of evolutionary history. Such fine-tuning, however, gives hostages to fortune. For when an environment changes, there is too little variation in the received phenotype for selection to operate on. A phenotype that maximizes fitness perfectly in one environment is so closely adapted to it that it may not retain enough variation to survive in any other environment.

Since we are the products of selection over changing environments, we are only approximate fitness-maximizers. Nature has produced us by selecting from what was immediately available for nature to shape to insure short-term survival. Doubtless, in its impatience nature has nipped in the bud potential improvements in our own species and in its predecessors. For the moment the only moral of this part of the story, for moral philosophy, is this: merely showing that altruism or other well-established patterns of morally praiseworthy action are strictly incompatible with monomaniacal, perfect, complete fitness-maximization is a poor argument for the claim that human behavior has become exempt from evolutionary selection. For we are not perfect fitness-maximizers. Natural selection has shaped us for *only approximate* fitness-maximization in the environments in which *Homo sapiens* has evolved. Approximate fitness-maximization leaves a great deal of room for non-adaptive altruism and other selfless actions.

3 PARAMETERS, STRATEGIES, AND THE MAXIMIZATION OF FITNESS

Now, among approximate fitness-maximizers, what sort of social behavior should evolve? This is a problem that arises with the advent of selection for living in family groups, which of course obtained long before *Homo sapiens* emerged. Until this point fitness-maximization is, in the game theorist's terms, "parametric," not "strategic." Which behavior is maximizing depends only on the environment, which provides parameters fixed *independently* of which behavior the organism is going to emit. But when organisms interact, which behavior one emits may be a function of what the other is going to do. So which behavior is fitness-maximizing will depend on how other organisms behave. This means that the optimal behavior is one that reflects a *strategy*, which takes account of the prospective behavior of other organisms. When social interaction emerges, fitness-maximization becomes a strategic problem.

This does not mean that, once groups emerge, organisms begin to calculate and select strategies based on recognition of the strategies of other organisms. It means something much less implausible. It means that those behaviors will emerge as fitter which, as a matter of fact, are coordinated with one another in the way they would have been, had they been the result of reflection and deliberation. This is because there is enough time for fitness differences between the rarer *but fortuitously* coordinated behavioral phenotypes and more common uncoordinated ones to pile up and select the coordinated ones.

Coordinated behaviors are sometimes cooperative ones; they are other-regarding, involving putting oneself at the mercy (or, at least, at the advantage) of another. Thus, they constitute a significant component of morality. The emergence of coordinated behaviors makes one sort of scenario for the emergence of morality tempting. This is the *group selection* scenario, according to which nature selected societies and groups because their institutions, including their moral rules, are more adaptive for the group as a whole. On this model, selection proceeds at the level of the individual (for individual fitness-maximization) and at the level of the group (for group survival and growth). The idea that evolution might lead to the emergence of morality by selecting for groups that manifest moral rules and against groups that retain a state of nature is, on the face of it, more attractive than trying to find a story of how morality might have emerged at the level of the individual. For the fitness-maximizing individual is concerned only with maximizing its offspring; it is ready to sacrifice others to this end. The fact that there are so many immoral and amoral people around makes implausible the notion that morality emerged among *Homo sapiens* the way opposable thumbs did – as an individual response to the selection of individual organisms. But the emergence of morality as a group institution is at least compatible with the observed degree of moral imperfection among individuals. Selection at the level of the group does not require uniformity among the individuals who compose it; no championship team has ever had the best players in the league at every (or even any) position (cf. New York Mets, 1969).

So, one way to reconcile other-regarding cooperative behavior with monomaniacal evolutionary egoism is to locate selection for cooperative institutions at the level of the group and selection for individual fitness-maximization at the level of the individual. If the forces selecting for the adaptation of groups are independent of those selecting for the adaptation of individuals, then those groups within which cooperation, promise-keeping, property, fidelity, etc., emerged, for whatever reason, might do better, last longer, or have larger, healthier populations than those groups which lacked such virtuous institutions. Thus, morality is explained as an evolved holistic social constraint on individual selfishness.

This is a nice idea, but one which evolutionary biology must exclude. For no matter how much better off a society with ethical institutions might be than one without them, such a society is seriously unstable, and in the evolutionary long haul must fall victim to its own niceness. The reason is that selection at the level of the group and the level of the individual are never independent enough to allow for the long-term persistence of a moral majority and an immoral minority. In fact, they aren't independent at all. The latter will eventually swamp the former.

Consider a society of perfectly cooperative altruistic organisms, genetically programmed never to lie, cheat, steal, rape, or kill, but in which provisions for detection and elimination of organisms who do not behave in this manner are highly imperfect (as in our own society). Since everyone is perfectly cooperative, the society needs no such provision. Now suppose that a genetically programmed scoundrel emerges within this society (never mind how – it might be through mutation recombination, immigration, etc.). By lying, cheating, stealing, raping, and otherwise free-riding whenever possible (recall that the detection and enforcement mechanisms are imperfect), the scoundrel does far better than anyone else, both in terms of well-being, and in terms of eventual fitness-maximization. He leaves more offspring than anyone else. If his antisocial proclivities are hereditary, then in the long run his offspring will come to predominate in the society. Eventually, 100 percent of its membership will be composed of scoundrels and its character as a cooperative group will long since have disappeared.

Evolutionary game theorists have provided a useful jargon to describe this scenario: a group with a morally desirable other-regarding strategy is not "evolutionarily stable": left alone, it will persist, but it can be "invaded" even by a small number of egoists – who will eventually overwhelm it and convert the society into one bereft of other-regarding patterns of interaction. By contrast, a society composed wholly of fitness-maximizing egoists is an "evolutionarily stable" one: a group of such egoists cannot be successfully invaded by some other, potentially nicer pattern of behavior. Its members will all, one after another, play the nice guys for suckers and out-breed and ultimately extinguish them.

The trouble, then, with group-selectionist explanations of the emergence of morality is that a group of other-regarders might do better than a group of selfish egoists, but it is vulnerable to invasion by one such egoist, an invasion which evolutionary theory tells us must always eventually occur – since nature is always culling for improvements in individual fitness-maximization. Whether from within or without, scoundrels will eventually emerge to put an end to other-regarding groups by converting them into societies of fitness-maximizers.

If morality is to emerge from the nature of organisms as approximate fitness-maximizers, it will have to happen at the level of individual selection. And it will have to be selection for optimizing behavior in the context of "strategic" interaction, where the optimum behavior of each organism depends on the behavior of other organisms. The trouble is that game theorists have increasingly come to suspect that there is no optimal strategy under these circumstances. If this is right, then there is none for evolution to choose, and no way for moral institutions to evolve from the strategic interactions of fitness-maximizers.

The problem of an optimal strategy for nature to select is easily illustrated in the children's game of Rock, Paper and Scissors. In this game, kids pick one of the three choices. Rock breaks Scissors and so beats it, Scissors cut Paper and so beat it, but Paper covers Rock and so beats it. Whether your choice wins depends on what the other kid picked, and no choice is better than any other. In an evolutionary situation like this, no strategy ever comes to predominate. Of course, if you know what your competitor will pick, you can always win. But what the other kid picks is going to depend on what he thinks you will pick. So, you have to know what he thinks you will pick in order to pick the best strategy, and so on backwards *ad infinitum*. There is no end to the calculation problem, and therefore no optimal strategy in the Rock-Paper-Scissors game. Game theorists have labeled the problem of this sort of game with no finite solution – no best strategy for any player – "the problem of common knowledge." In principle, the problem of having to infinitely iterate calculations about what other players will do bedevils most strategic games.

While the problem of common knowledge cannot affect organisms which are incapable of making calculations about the strategies of others, it can affect the evolution of fitness-maximizing strategies. As nature selects the best among competing strategies for fitness-maximization, it must eventually face contexts in which the best strategy for an organism to play depends on what other strategies are available to be played by other organisms. If the game theorist can prove that, in the long run, there is no single best strategy – even with rational calculation on the part of the players – then we can expect natural selection to do no more than produce a motley of equally good or bad strategies that compete with one another, at best gaining temporary ascendancy in a random sequence. In other words, natural selection will produce nothing but noise, disorder, no real pattern in the behavior of fitness-maximizers who face strategic competition as opposed to parametric optimization problems.

It seems safe to assume that *Homo sapiens* has not in fact suffered this fate. For the most part, our interactions do show a pattern, and an other-regarding, cooperative one at that. Morality is the rule and not the exception (and not just one among a series of cyclically succeeding patterns of behavior). It must follow, therefore, that evolution has not led us (or our evolutionary forebears) down the cul-de-sac of the problem of common knowledge. But, if game theorists are right, almost the only way evolution could have avoided this sort of chaos is for other-regarding principles of conduct to have emerged in parametric contexts, and then to be evolutionarily stable, un-invadable when these contexts became strategic.

4 KIN-SELECTION AND UNCERTAINTY

For a fitness-maximizing organism, interactions with offspring are close to being parametric. For, almost no matter what children and kin do to you, if you act in their interests, the result will increase your fitness. The fitness-maximizing strategy for an organism is therefore to act so as to maximize the fitness of its offspring. Thus, in selecting for fitness-maximization, nature will encourage organisms whose genetically encoded dispositions include sharing and cooperating, and even unreciprocated altruism towards kin – children, siblings, even parents. For these strategies are likely to increase one's offspring, no matter how they respond to you. This sort of kin-altruism is evolutionarily stable and un-invadable. A short-sighted, selfish organism, who behaves as though its own survival or that of its children counted for its fitness, would end up with fewer offspring over the long haul. For sometimes it would look out for number one (or number one's kids) when sacrificing itself or a child would result in the survival of a larger number of offspring (recall, that under sexual reproduction, a child is only half an offspring). In selecting for fitness, nature will select for "inclusive fitness" and "kin-selection" will emerge. Kin-selection is something we can count on emerging long *before Homo sapiens* appears. It becomes an adaptive strategy as soon as the number of genetic offspring begins to exceed the number of children. (Recall, three nephews carry more of an individual's genes than one child.)

When *Homo sapiens* emerges, therefore, we are already beyond Hobbes's state of nature. Cooperation, altruism, and other-regarding behavior generally is already established inside both the nuclear and the extended family. Indeed, it is likely to already have been established a bit beyond this. Consider that individuals do not wear name tags or carry their genealogy on their sleeves for others to examine before deciding whether an interaction will be parametric or strategic. There are, of course, clear signs of kinship that even animals with limited recognition powers can use: odor, proximity to a nest or region. And there are clear signs of xenonimity – strangerness. But there is always a large area of uncertainty in between, a range of interactions in which two organisms just can't tell with any more than a certain moderate level of probability whether they are kin or not. This will be more true for males and their putative offspring than for females. Given the nature of procreation and gestation in many mammalian species – and especially in *Homo sapiens* – the male can never be as certain as the female that the young in his family are his offspring – i.e., that they share some of his genes. Unless the female is under constant and perfect surveillance during the critical period, the question of whose sperm fertilized her ovum must always be a matter of some doubt. Beyond the relation between mother and child, the degree of consanguinity between any two organisms is always a matter of probabilities, and doubts about kinship are easier to raise than to allay.

Under conditions of uncertainty about kinship, what is the optimal strategy for a fitness-maximizer? Game theory tells us that the rational thing to do is to apportion the degree of one's other-regarding behavior to the strength of the evidence of consanguinity. In the long run, as natural selection operates, it must favor this strategy as well. Even in cases where the available positive evidence of consanguinity (subtle similarities of smell, coat, color, shape of beak, pitch of mating call, etc.) is difficult to detect, one can expect nature to select for cooperation and other-regarding behavior between kin, provided only that it has enough time to fine-tune the detection mechanisms. Considering the job it has done in optimizing the eye for vision within the time constraint of four million years, it may seem reasonable to suppose it can fine-tune kin-selection strategies as well. And if everyone turns out probably to be closely enough related to everyone else, then natural selection might be expected by itself to produce other-regarding behavior up to levels of frequency that match the probability of universal consanguinity. Here we have the emergence of morality, or at least a crucial aspect of it, without having to solve the problems common knowledge makes for strategic games.

However, the amount of other-regarding behavior that might in fact be fitness-maximizing just because of the fact that we are all each other's seventh cousin hardly seems sufficient to explain the emergence and persistence of moral conduct. The problem with this neat explanation is that we have no independent idea of whether the payoffs (in more offspring) for being other-regarding are really great enough when the probability of being related falls to the level that obtains between you and me. And there doesn't seem to be any easy way to find out. In short, our explanation isn't robust enough. It rests on a certain variable taking on a very limited range of values, one within which we have no reason to think it falls. We need a better explanation for the emergence and persistence of other-regarding behavior than kin-selection and the uncertainty of relatedness can give us. It's all right to start with kin-selection, but we need an explanation that carries other-regarding conduct into the realm of strategic interactions among fitness-maximizers *unlikely* to be kin.

To do this, we need to help ourselves to another brace of healthy assumptions about morality and game theory. First, let us accept without argument that the institutions of morality are public goods: they cannot be provided to one consumer without being provided to all others, so that any one consumer has an incentive to understate the value of the good to him and so decline to pay its full value provided he is confident that others will pay enough to provide it. Certainly, the institution of generalized cooperation is like this. No one can count on it unless everyone can, and we all have an incentive to understate its value to us whenever we are asked to pay our fair share to maintain it. Moreover, fitness-maximizers have an incentive to cheat, to decline to cooperate, if they can get away with it undetected or unpunished. But if everyone knows this, and everyone knows that everyone knows this, etc., then the institution of cooperation will break down because of our common knowledge. The public good is lost, and every one is worse off. The prisoner's dilemma graphically illustrates this problem of the provision of public goods. Individual rational agents have an incentive to be free-riders, to decline to cooperate. The result is a non-optimal equilibrium in which no cooperation is visible. The natural selection version of this collapse from the fortuitous provision of public goods to a non-optimal equilibrium takes time, as individual defectors emerge through recombination or mutation and out-reproduce cooperators.

The second assumption we need is that most of our morally relevant interactions are moves in an indefinitely long sequence of prisoner's dilemma games. This seems a not unreasonable assumption: honoring moral obligations is not a one-shot, all-or-nothing affair. It is a matter of repeated interactions largely among the same individuals. Interactions with strangers are by definition less frequent than with people we have interacted with and will interact with in the future. Now, one important fruit of the joint research of game theorists and evolutionary biologists has been the conclusion that, even among strangers, being a free-rider (always declining to cooperate, always taking advantage) is not the fitness-maximizing strategy in an iterated prisoner's dilemma. Rather, the best strategy is what is known as "tit-for-tat": that is, for optimal results one should cooperate on the initial occasion for interaction, and on each subsequent occasion do what the other player did on the last round. This strategy will maximize fitness even when everyone knows that everyone else is employing this strategy. For even on the assumption that there is complete common knowledge of what strategies will be chosen, tit-for-tat remains the best strategy. Once in place, it assures cooperation even among unrelated fitness-maximizers. It circumvents the common-knowledge problem.

5   ETHICS – QUICK AND DIRTY

There is one rather serious obstacle to natural selection's helping itself to this strategy: the problem of getting it into place. For tit-for-tat cannot invade and overwhelm the strategy of narrow selfishness that is required by strict fitness-maximization. In a group of organisms that never cooperate, anyone playing tit-for-tat will be taken advantage of at least once by every other player. This advantage is enough to prevent tit-for-tat players eventually swamping selfishness. In fact, it may be enough of an advantage for tit-for-tat to be driven to extinction by the strategy of selfishness every time it appears as a strategy for interaction.

This is where nature's preference for the quick and dirty, *approximate* solution to the problem of selecting for fitness-maximization comes in. Our approximate fitness-maximizers' optimum strategy involves other-regarding behavior with kin, and selfishness with others. How will fitness be maximized in the borderline area where kinship and its absence are difficult or impossible to determine? When the only choice for an organism is to cooperate or decline to do so, how does it behave? By flipping a coin weighted to reflect the evidence for kinship, and doing as the coin indicates? A few pages back I derided this suggestion, though we cannot put it past nature to have evolved a device within us that has this effect. On the other hand, nature will prefer quick and dirty solutions to mathematically elegant ones, provided they are cheap to build, early to emerge, and do the job under a variety of circumstances, etc. If tit-for-tat is almost as good a strategy for fitness-maximization in cases of uncertainty as employing the probability calculus and far easier for nature to implement, then on initial encounters under uncertainty about kinship, individuals playing this strategy will cooperate. But this means they will cooperate thereafter as well. Thus, interactions at the borderline come to have the character of interactions within the family; parties to any and every interactive situation will generally cooperate.

Now suppose that among organisms genetically programmed to be other-regarding within the family and to play tit-for-tat at the borderlines, one or more individuals emerge with a new variation: their genome is programmed to encourage tit-for-tat always and everywhere, or at least whenever interacting with strangers. Interaction with selfish strangers will be costly to such organisms and should lead to their extinction. But suppose such interaction is rare. Furthermore, suppose (as seems reasonable) that the strategy of always playing tit-for-tat is otherwise an adaptive one, with advantages over other more complex strategies, especially for organisms lacking complex cognitive and calculational powers. For the cost of maintaining and using a storage system for kin and non-kin may be greater than the cost of being taken for a sucker in just the first round of an indefinitely iterated interaction. This will likely be true when the chances of meeting a stranger are extremely low, as they will be in the earlier stages of the evolution of mammalian species living in family groups.

What is more, as kin-groups are submerged in larger groups of agents who interact with one another more frequently than with others, these groups will in fact acquire the kind of cohesiveness that looks as much like group selection as kin-selection does: just as kin-selection in effect selects for the family-group by increasing the frequency of interaction-opportunities among individuals, larger social groups which keep agents playing with one another, or bring them back into more frequent interactions from time to time, will also increase the adaptational advantage of a tit-for-tat strategy. Such groups will be adaptationally advantaged to the degree their members are individually advantaged. Whence, in the view of some biologists and philosophers, group selection is not only possible but commonplace in human societies [cf. Sober and Wilson, 1998].

It is, in general, easy to imagine scenarios that make tit-for-tat the best overall strategy under most circumstances in a given environment. But this means that natural selection for approximate fitness-maximization among individuals has led to the emergence of cooperative, other-regarding strategies. It has solved the problem of providing public goods to individual organisms geared always and only to look out for themselves and their kin. If ethical institutions are, after all, public goods, then we have explained how they might emerge among approximate fitness-maximizers.

Of course, this entire story applies to us only to the extent that we are approximate fitness-maximizers. This is not hard to show. In fact, if anything, it's too easy to show. For the story does not include any indication of how good an approximation to perfect fitness-maximization is required for the emergence of other-regarding strategies. Even if it did, we have no idea of whether *Homo sapiens* is in fact a good enough fitness-maximizer for this scenario actually to obtain. For these reasons, the claim that we are in fact approximate fitness-maximizers will have vanishingly small empirical content. But then empirical content was never the strong point of any evolutionary theory, and is of little interest in moral philosophy anyway.

That we are fitness-maximizers to some degree of approximation goes without saying. After all, the only alternative to being an approximate fitness-maximizer is being extinct. And how did nature shape us for fitness-maximization? What phenotypical properties of *Homo sapiens* did it shape in this direction? Well, the quickest and dirtiest way of making us approach fitness-maximization is to make us *approximate utility-maximizers*, to shape us into systems organized to maximize our well-being, by linking well-being to the avoidance of discomfort, pain, and distress, and the attainment of comfort, pleasure, and feelings of security. The reason is obvious: an organism's reproductive potential is, *ceteris paribus*, a function of its well-being. So, in order to select for fitness-maximization, nature will select for organisms that by-and-large maximize their well-being. The by-and-large clause reflects the fact that there are certain departures from utility-maximization that nature will select for too. For example, it will select for organisms that sacrifice their own well-being to offspring, especially after they have passed the age of optimal procreation. Or, equivalently, nature will select for preference structures that make kin-altruism pleasing to the individual. This is a quick and dirty solution to the problem of programming kin-selection, one which corresponds to the philosopher's claim that altruism is just the reflection of a perverse preference structure.

If the quick and dirty solution to the problems of designing an approximate fitness-maximizer is to design an approximate utility-maximizer, then our merely plausible explanation for the emergence of morality or of one important component of it may have another role to play. It may turn out to be a part of a (weak) *justification* of morality, or at least of one important component of it.

One traditional question of interest to moral philosophers is that of how to convince the rational egoist to be moral, how to show the egoist that being moral is in his interest. Nowadays this problem is often set forth as that of showing how morality could be part of a strategy that maximizes individual utility. In its own way, natural selection provides reason to suppose that morality is part of a utility-maximizing strategy, and our story provides a plausible scenario for how this might have happened.

It is clear that nature began selecting for utility-maximizers long before it began selecting for other-regarding cooperators. For one thing, maximizing well-being is a strategy to be found across the phylogenetic spectrum; it doubtless characterized our ancestors long before the rearing of offspring in nuclear or extended families and the emergence of social groups made other-regarding cooperation possible and necessary. Having laid down very early in evolution approximate utility-maximization as a quick and dirty strategy for approximating fitness-maximization, nature is unlikely ever to "rip it out" and start over. This means that when it lays down other strategies, they will at least have to be compatible with utility-maximizing. It's much more likely that the new strategy will be new ways to maximize utility under new circumstances. But if cooperation and other-regarding behavior generally is nature's way of most efficiently maximizing utility, then it should be good enough for us. That is, in our own calculations and reflection on how to maximize our utilities, we should expect to come eventually to the same conclusion which it has taken nature several geological epochs to arrive at. Both rational agents and nature operate in accordance with principles of instrumental rationality; they both seek the most efficient means to their ends. Since nature's end (approximate fitness-maximization) is served by our ends (approximate utility-maximization), our means and nature's will often coincide.

6 CONCLUSION

It's a nice story, and it seems to have a moral for moral philosophy. I think it is absolutely the best biology can do by way of shedding light on anything worth calling "human nature" and drawing out its implications for matters of interest to moral philosophers. But before taking any comfort in it at all, we need to recall and weigh the hostages to fortune it leaves – the many special assumptions about us and about the nature of moral conduct that it requires just to get off the ground: to begin with, the idea that just because one is cooperative or other-regarding, one has attained the status of a moral agent or some important precondition to it. Then there are the claims about humankind as approximate fitness-maximizers. Even if you accept this view of "human nature," as I do, you are committed to a level of fitness-maximization that you cannot specify beyond saying it is high enough to allow for the scenario I have tried to unfold. Then you have to find a way to draw the force or circumvent the difficulty of the problem of strategic games, in which there seem to be no stable equilibria in the behavior of fitness-maximizers, let alone equilibria that underwrite any part of morality. (And you can't call upon group selection to help solve this problem.) Then you have to buy into the theory of kin-selection and its application to conditions of uncertainty. This is one of the smaller gnats to strain at, given the independent evolutionary evidence for kin-selection. But the trouble is that it will not suffice when interactions begin to transcend the family. At this point, we need to assimilate morality further to strategies of choice to be analyzed by the tools of economics and game theory. Finally, we need to be able to fudge our account enough to say that morality emerges because we are not perfect fitness-maximizers, since the best nature can do is make us approximate utility-maximizers.

But perhaps the most difficult consequence of this story to swallow is this: if nature had been able to do any better, morality might never have emerged at all.

NOTE

Reprinted with permission from *Social Philosophy and Policy*, 8:1 (Autumn 1990): 86–101.

7

Moral Realism and Social Science

The traditional intersection between the philosophy of social science and moral philosophy has been the problem of value-freedom: whether theories in social science unavoidably reflect normative judgments, whether they can be or are free of them, and whether either of these alternatives is a good thing or not, given the aims and methods of empirical science in general. Despite its importance, this is a debate to which little both novel and reasonable has recently been added.

But the recent resurgence of interest in "moral realism" holds out the hope of a new approach to this issue, one likely to lend support to the opponents of moral neutrality, support from a direction in which they could not have expected it – the "scientistic" approach to theories about human behavior. For once moral realism is embraced, the attractiveness of deriving moral claims from descriptive social theories becomes very great.

1 REALISM AND NATURALISM

Moral realism is not just the thesis that at least some normative claims are definitely true or false. It also requires that at least some of them are in fact true. A moral realist who held that all ethical claims are just false could hardly be said to have offered a defense of ethical theories and claims. But if at least some moral claims are true, it behooves the moral realist to show how we can know which ones they are, and what the evidence for them is. But providing the metaphysics and epistemology moral realism needs has long been a problem. As Sayre-McCord [1988, p. 13] notes, "the common (mistaken) assumption is that the only realist positions available in ethics are those that embrace supernatural properties and special powers of moral intuition." The assumption is said to be mistaken because another "ism" – naturalism – has again become fashionable in meta-ethics. This is the brace of theses that (a) the conditions that make some moral claims true are facts about the world and its denizens, ontologically no different from the facts dealt with in physics or psychology, and (b) the way in which we can come to know such claims to be true is identical to the ways in which scientific claims in general are acquired – by the formulation and confirmation of theories that explain our observations.[1] To be plausible, moral realism needs to avoid any tincture of ethical intuitionism or metaphysical mystery mongering. Naturalism is the only option available to realism for avoiding the charge that its metaphysical and epistemological foundations are untenable.

This is where the social sciences and the philosophy of science come in. The relevant observations against which to test explanatory theories embodying moral claims must be observations of human behavior, individual and aggregate, and both the theories and the observations are the province of the social sciences.[2] If the case for social scientific theories embodying normative claims about the good, for example, can be made out, then not only will moral realism be vindicated in principle, but the normative bearing of social theory, independent of any further evaluative premises, will also be established. This would be decisive refutation of the thesis that social science, like natural science, can be or should be value-neutral. No one currently writing on moral realism, and the social-scientific naturalism it requires, has noticed this special "dividend" of their theory. Nor have they noticed that a decisive argument for *vert-frei* social science will *ipso facto* be an argument against naturalism and moral realism.

Proponents of value-free social science and opponents of moral realism thus have common cause. Regrettably, they can only pursue their cause in a case-by-case piecemeal way. A general impossibility proof that no empirically adequate social theory that both explains the observations we make about human affairs and embodies moral claims is forever unavailable. For providing such a proof is equivalent to establishing a negative existential claim. This means that the only way to defend anti-realism in ethics and moral neutrality in social science is to attack the individual particular theories that are advanced as both empirically adequate and morally significant.

Surprisingly, though much has been written on the generic subject of moral realism and naturalism, few such theories have even been sketched. This is surprising since without such details, moral realism-*cum*-naturalism remains at best a pious hope, a bare logical possibility, instead of a robust foundation for ethics, a bulwark against ethical relativism, moral skepticism, and social nihilism.

2 RAILTON'S PROJECT

One recent attempt to fill out the details of the requisite social theory is that of Peter Railton. In "Moral Realism" [Railton, 1986] he offers an argument for moral realism that recognizes the obligation to say what the explanatory theory naturalism demands would look like. In the rest of this essay, I examine this theory and Railton's arguments for it. My doubts about it stem as much from reflection on the importance of value-neutrality in adequate social theory as they do from doubts about notions that moral claims can be true or false. My strategy is mainly to focus on the substantive and the methodological defects of the social theory Railton adopts to ground his moral realism. But some of my arguments reflect traditional objections to moral realism and naturalism that go back to Moore and Hume. At a minimum, my objections will reflect strictures on subsequent theories in social science for which normative force is claimed. But perhaps some of the problems to be noted are unavoidable defects in any theory that aims to combine empirical adequacy and moral force. If so, these objections may help undermine the whole project of moral realism.

When we see what is wrong with Railton's theory and what is revealed by a wider view of social phenomena than Railton's, there remains little reason to endorse either his positive social theory or the integrally associated normative theory of moral goodness. And when repaired to avoid these defects, the result is either a positive theory without ethical ramifications, or one quite irreconcilable with strongly held moral convictions of the sort Railton seeks to justify and explain. Is Railton's the best-case argument for moral realism from social sciences? From my perspective, it must come close to being so. For Railton's argument is self-consciously naturalistic and avowedly Darwinian. Moreover, I share the moral values Railton hopes to extract from social science. If Railton does not succeed, I cannot be optimistic about anyone else's success in this enterprise.

Railton is a naturalist. He sets himself the task of showing that "moral facts are constituted by natural facts" – that is, facts with an explanatory role in respect of human behavior. To do this, he writes, moral facts must exhibit two features: (1) existence independent from our own beliefs and a determinate character independent of these beliefs; and (2) it must be possible for these facts causally to influence our mentation and behavior (p. 172). These features, of course, do not make facts moral, they make them "real." That facts are moral consists in their instrumental rationality "from a social point of view, as opposed to the point of view of any individual" (pp. 190, 191). Thus, as Railton eventually notes, moral goodness will turn out to be consequentialist, interpersonally aggregative and maximizing (pp. 190–91, note 31). It will be what best serves the interests of all members of society somehow taken together. This approach to moral goodness already commits Railton to an empirical social theory that is both holistic and functional, as we shall see. It means that his moral theory will bear both the strengths and the weaknesses of such theories. In this respect Railton's approach to moral realism is typical of attempts to combine ethical and empirical social theory (though his attempt is far more detailed than others).[3]

3   REALISM ABOUT NON-MORAL VALUE

Railton's strategy is to produce an account of what he calls non-moral value, and then to parlay this theory into an account of moral value, showing that the latter has all the reality of the former.

What is non-morally valuable to an agent is not what he actually values, but what he would value if he knew his "real" interests and employed this information rationally. Individuals are often notoriously wrong about their real interests, mainly as a result of epistemic and logical errors of omission and commission. Consider an agent with a certain set of preferences. Now deprive him of all of his false beliefs about himself, his situation, the consequences of acting on these preferences, and add all relevant knowledge about himself, his situation, and his preferences. The preferences that remain, and those that are added by this transformation of his doxastic states, are his "real" or "objective" interests. These interests obtain in fact and for all of us, whether we know it or not. Subjective preferences that conflict with them are not really in the interest of the agent no matter how strongly he embraces them. Now, some state of affairs is "non-morally good" for an agent if it would satisfy an objective interest of the agent – "roughly, what he would want himself to seek if he knew what he was doing" (p. 176). The non-morally good thus turns out to "correlate well with what would permit the agent . . . to experience physical or psychological well-being. Surely our well- or ill-being are among the things that matter to us most, and most reliably, even on reflection" (p. 179). Thus, our well-being is intrinsically non-morally good, and whatever is instrumental in achieving this good is an "objective interest."

This account makes Railton's a thoroughly biological theory of the "non-morally good," perhaps more biological than he realizes. For interests are objective if they enhance well-being, and well-being is an intrinsic good for us because natural selection made us that way. It did so because well-being is causally connected to enhanced reproductive rates, and so secures evolutionary adaptation. This means that well-being is an intrinsic good to individuals because it is an instrumental good to the lines of descent in which they figure. Given environmental circumstances, certain of our ends are conducive to the demographic expansion (or at least the persistence) of each of our respective lineages. So, evolution has shaped us to treat them as objective interests. Railton's evolutionary explanation for our objective interests is consigned to a longish footnote (p. 179). But its importance is evident. Railton's naturalism requires that non-moral goods be grounded in biological facts; his instrumentalism about rationality precludes treating well-being as an intrinsic good, *tout court*.

Like any adaptation, an interest is objective only with respect to an environment. When an environment changes, an interest can go from objective to subjective, from adaptive to maladaptive. For example, in the time of the hunter-gatherer, eating as much protein as possible was adaptive and in the objective interests of agents. Today in the West, it is highly maladaptive, and not in the objective interests of agents. Why? Because the environment has changed the amount of protein available to many people from barely sufficient to practically unlimited.

It is crucial to see that a theory of objective interests, of what conduces to well-being, must leave room for the environment as a variable: x is an objective interest for A in environment e. Railton does not omit mention of the environment in the exposition of non-moral goodness: he refers to the agent's circumstances. But he does omit it in his theory of moral goodness, with serious consequences.

Railton believes the notion of "objective interests" may yield counterintuitive results. In particular, he holds, it may turn out that quite aberrant and indeed repulsive preferences will figure among the objective interests of an individual, and thus be part of his "non-moral good." He writes, "some people . . . might be put together in a way that makes some not-very-appetizing things essential to their flourishing, and we do not want to be guilty of wishful thinking on this score" (p. 177). On the other hand Railton may wish to take some comfort in the "pluralism" about objective interests this possibility promotes. For few will want to adopt an account of the non-morally good that allows for little interpersonal variation in objective interests. But in fact, whether counterintuitive or not, this variability is not something Railton's theory really allows him. Because of its unavoidable biological character, Railton's conception of "objective interest" is more likely to homogenize agents than it is to reveal their divergent objective interests. To the extent that our interests – both objective and subjective – have what Railton calls a "reduction base" in our physiology and environment, acquiring complete knowledge about ourselves includes acquisition of information about this base as well. If, as seems reasonable, sadism, for example, turns out to be symptomatic of a physiological malfunction, then it will be in the objective interests of a sadist to treat the disorder. Similarly for more innocent pleasures. The insatiable craving for sweets, when viewed in the light of all its long-term effects, can be no part of an objective interest either.

This is true unless, of course, mental and physical malfunction can enhance "physical or psychological well-being." If so, a disorder or illness may figure in the objective interests of an agent. But it would be difficult for Railton to allow that there are some people whose objective interests are directly, intrinsically served by the exercise of cruelty or the onset of dental pain – unless our objective interests were somehow decoupled from evolutionary constraints, as indeed in modern life they sometimes seem to be. But in this case they would be decoupled from well-being and lose their standing as non-morally good. Such maladaptive objective interests might still be able to explain behavior, especially aberrant actions, as rational (in Railton's instrumental sense). But there would be no reason to identify such maladaptive interests as non-morally good. For a naturalistic approach would identify them as pathologies. Cut off from evolutionary adaptation, such interests no longer have the kind of naturalistic explanation in terms of adaptation that Railton's realism requires. Their naturalistic explanation is a matter of "breakdown" or "malfunction." An interest which is the result of some malfunction, or deficiency, is indistinguishable from what Railton calls a subjective interest, one which may be the product of epistemic and logical errors of omission and commission, for, after all, these are malfunctions or deficiencies themselves. So aberrant objective interests are not really explained on Railton's theory as non-moral goods. And its prospects of accommodating the wide variation in human aims and objectives are not as great as Railton hopes.

In and by itself there is nothing philosophically very objectionable about Railton's theory of the non-morally good. It is, as noted, rather more biological than he indicates and may undercut conventional views about the degree to which individuals can differ in their real interests. But perhaps its gravest defect is not a philosophical one at all. Railton bills the account of "objective interest" as part of an explanatory theory of human behavior. But if there is such a theory, it is not a very interesting one. For consider, "objective interests" help explain agents' behaviors because the subjective desires that lead to action are supposed to evolve in the direction of these objective interests:

> [An individual's] desires evolve through experience to conform more closely to what is good for him, in the naturalistic sense intended here his receipt of ideal

information, but rather of largely unreflective experimentation, accompanied by positive and negative associations and reinforcements. There is no guarantee that the desires "learned" through such feedback will accurately or completely reflect an individual's good. (p. 180)

Without the caveat of the last sentence this theory would be stigmatized as both Panglossian in its expectations about desires and patently false. On the other hand, the caveat deprives it of much explanatory power. For so qualified, the theory is consistent with the persistence of any kind of desires, no matter how non-morally bad, over any but the long run. In terms Railton employs later in his essay, the theory is guilty of a "complacent functionalism." Like another such theory, folk psychology, at best its explanations are highly generic or inevitably *ex post*. This fact bedevils the account of moral goodness which Railton wants to build on the model of non-moral goodness. It is to this account that I now turn.

4 MORAL GOODNESS

In line with the adequacy conditions set out at the beginning of "Moral Realism" (p. 172) Railton notes that to establish moral realism one must show that there are facts about what ought to be the case, and that these facts can be explanatory (p. 185). For facts about non-moral goodness, this should not be difficult. Hypothetical imperatives about what one ought to do in order to attain well-being, may be said to have "derived" explanatory force for they rest on declarative statements about what in fact conduces to well-being, and on the presumption that agents seek well-being. Railton seems to adopt the view that 'A does x' can be explained by 'A ought to do x' because this latter can be unpacked into "Creatures like A have evolved in ways that make doing x conducive to their well-being, and they seek their well-being." (Railton's argument for the explanatory power of normative statements proceeds in a different way, exploiting examples to make the view plausible [pp. 185–187], but I think the sketch given above fairly represents the foundations of his reasoning.)

It is worth nothing that a categorial imperative will not have the sort of explanatory power a hypothetical one does. 'A does x' cannot be explained by 'A ought to do x, without qualification or condition'. On the other hand, a categorial imperative is not vulnerable to the questions that a hypothetical imperative always leaves open: If A ought to do x because he seeks end e, then why does he seek end e, and why should he seek it? In the case of non-moral goods and individual agents, Railton's answers seem to be given in terms of the evolutionary adaptiveness of individual well-being and the further end of the survival of the individual's lineage. What will the answer be in the case of moral goodness and social aggregates? Any answer to this question suggests that moral goodness is instrumental to some further end. Since moral goodness is part of an explanatory theory, all the imperatives it sustains will be hypothetical and will generate the question of why does the system which attains moral goodness do so and why should it do so? (I take up this point further in section 6 below.)

Given his account of normativeness in terms of hypothetical imperatives with derived explanatory power, the steps Railton takes to an account of moral goodness and moral norms with explanatory power are pretty direct.

The non-morally good reflects what it is rational for an individual to want – his objective interests – in order to secure his individual well-being. The morally good reflects what it is rational to want, not from an individual point of view, but from "the social point of view" (p. 180). What is rational from the social point of view is "what would be rationally approved of were the [objective] interests of all potentially affected individuals counted equally . . ." So, moral goodness is "what is rational from the social point of view with regard to the realization of intrinsic non-moral goodness." Moral goodness is thus supposed to involve increasing some aggregation of the non-moral goodness of individuals, the well-being of agents in society – not just any aggregation, such as average expected utility or well-being, but an aggregation that takes account of all affected individuals equally. Railton does not indicate which aggregation will do this, so, hereafter, I will refer to social rationality as requiring some aggregation, meaning a particular one, and leaving it open, along with Railton, which one.

Non-moral goodness pertains to individuals, moral goodness pertains to "social arrangements – a form of production, a social or political hierarchy, etc." When such arrangements depart "from social rationality by significantly discounting the interests of a particular group" there is "potential for dissatisfaction and unrest" – and therefore reduction in the viability of the "social arrangement," and of the whole society which it characterizes. Although he does not say it explicitly, a moral injunction will be a hypothetical imperative in which the end served by maximizing some aggregation of the well-being of the society's members is that society's well-being or survival. Thus there will have to be a social theory identifying the prerequisites for the viability of well-being of the society, and the conditions that increase and reduce it. Otherwise there is no end for social rationality to serve. But as Railton notes, rationality is always instrumental.

Railton is not shy about endorsing such a theory, for he needs it to establish moral realism as a claim about the causal and explanatory force of independently existing moral facts. The theory proceeds as follows: an optimally rational social arrangement minimizes social unrest and discontent. Departures from it lead to "alienation, loss of morale, decline in the effectiveness of authority . . . potential for unrest, . . . a tendency towards religious or ideological doctrines, or towards certain forms of repressive apparatus, . . ." etc. (p. 192). These conditions are not good for the society or for the social arrangements that lead to discontent. That is, as such departures from social rationality increase, the prospects for these arrangements' continued existence decline. Since those social institutions which have survived are ones which have been selected for survival-promoting features – i.e., social rationality – it follows that social rationality, like any adaptation, has an explanatory role to play in social theory. Or at least it does to the extent that the theory is a functional or teleological one, assigning either some stable equilibrium to social processes, or some end-state towards which societies evolve, given their environments. (As noted below, however, Railton seems to eschew the notion that a teleological account of social arrangements presupposes some kind of stable equilibrium. [Cf. p. 194.] This is simply incompatible with the adoption of a functionalist feedback theory. A teleological process must track some variable over a range of conditions. So long as it keeps tracking, it is either at equilibrium or moving towards it and showing stability.)

Railton believes that recent work in social history and historical sociology tends to confirm such a theory. This work explains observed social change as movement in the direction of greater social rationality through a feedback between changes in the level of discontent and fuller recognition of the interests of social groups. Of course, Railton admits, actual short-term trends sometimes move away from social rationality, but the discontent this breeds tends to move society back in the "right" direction, one which tracks improvements in social rationality. Thus, social rationality constitutes a fact about moral goodness open to empirical study, and norms asserting the obligation to enhance social rationality can have some explanatory power. Besides endorsing particular works in social science that reflect this theory, Railton cites three gross trends, which he believes substantiate it: (1) "Generality" – the decline of tribalism, ethnic exclusivity, ethnocentrism, and xenophobia; (2) "Humanization" – the demystification of ethics and its separation from religious foundations; and (3) "Patterns of Variation" – greater approximations to social rationality in cases where such arrangements are more obviously in the subjective interests of all participants (p. 198).

I do not share Railton's confidence in the social science he cites, and I find at least two of his trends susceptible of interpretations which make them appear to be movements away from "social rationality." Thus, the first could equally well be described as the overwhelming of cultural differences in the interests of some covert or overt imperialist imperative; certainly the socially enforced abrogation of traditional ways of life has led to considerable discontent (cf. Iran under the Pahlavi dynasty), and could therefore be construed as a departure from social rationality. (Compare the biological irrationality of extinguishing species which provide a reservoir of variations we may need hereafter.) The "humanization," in Railton's terms, of moral prescriptions has certainly reduced their subjective force. When people thought that morality was God's command, they were more likely to embrace it. To the extent moral conduct is socially rational, this trend, too, seems in the opposite direction. Finally, in cases where "social rationality" can be explained in terms of individual rationality, it is in effect explained away, made superfluous as an explanatory concept. Naturally, there are possible counters to each of the objections briefly sketched here. But even if the gross historical trends are as Railton makes out, there is something crucial that is missing from his theory. Railton says that his theory is no mere "complacent functionalism or an overall endorsement of current moral practice or norms." He excuses himself from this charge by pointing out that the account "emphasizes conflict rather than equilibrium and provides means for criticizing certain moral practices..." (p. 199). But conflict and equilibrium are by no means incompatible in a functionalist theory: indeed many such theories (including "complacent" ones, like "democratic pluralism") secure equilibrium through conflict. Consider game theoretical models of evolutionary interaction. And it is by no means clear that Railton's theory of social rationality provides the resources to mount serious criticism of any very long-standing social practice, present or past, as we shall see.

More seriously, Railton's theory is a complacent one, almost by default – because it fails to specify the environmental factors against which social institutions measure up for rationality or adaptiveness, it fails to indicate how they have changed over time and fails to indicate how these changes shift social arrangements from the adaptive to the maladaptive, the rational to the irrational, and back again. Instead he simply assumes that the background conditions always and uniformly favor increasing the aggregate well-being of all agents treated equally as socially rational.

To see how this assumption vitiates Railton's theory, we need to answer the question of for whose ends (and which ones) it is rational to optimize some aggregate of individual well-being. There are only two possibilities:

optimizing aggregate well-being either serves the ends of individuals or of the society composed of them. Neither of these alternatives are satisfactory. Treating the ends of society as a whole, however, has the virtue of not making the theory empirically false.

Optimizing some aggregate of individual well-being is rational either for society as a whole, treated as an individual, or for the individual agents who compose it. There is no third possibility, since that is all there are: individuals and the aggregate of them. But clearly optimizing average, or median, or minimum, or maximum well-being or any such social welfare measure is not rational for the objective interests of every individual member of society. Some will be disadvantaged no matter what measure of aggregate well-being is optimized. In fact, under some circumstances of scarcity, for instance, that have been realized in the course of human history, the rational course may involve sacrificing the objective interests of a majority of the members of a society.

If social arrangements turn out to be morally good if and only if they are non-morally good, through some sort of disaggregation, for all the individual members of society, then of course there are almost no social arrangements that are morally good, and none that we could devise that would be. And if social arrangements are morally good because they are non-morally good for some subset, perhaps even a very large subset, of the individuals who compose a society, then we need a naturalistic explanation of what makes this subset the morally relevant one. For example, an evolutionary argument that the well-being of this subset is necessary or sufficient in most circumstances for the well-being of the society as a whole, or one that held its well-being was equally in the objective interest of every member of the society individually.

Railton does not give us such an argument. But his discussion of moral principles as "non-indexical" and "comprehensive," as reflecting "a social point of view" (pp. 189–90), suggests that social arrangements are rational to the degree they serve the ends of the whole society, as distinct from the individuals who compose it. This is not a view Railton will accept, for he holds that the only bearers of value are actual sentient beings, but it is the only way to reconcile his instrumental rationality with the maximization of aggregate well-being. So, in what follows let us explore the implications of this answer to the question of whose ends aggregate well-being is instrumentally rational for: it is instrumental for the ends of society as a whole.

Social rationality, like evolutionary adaptation, or individual rationality for that matter, is always relative to an environment. What is adaptive with respect to one environment is maladaptive with respect to another. Consider what happens to polar bears that move south. At a minimum, Railton needs to say that constant environmental factors make some aggregate maximizing of well-being an adaptation of all societies everywhere and always: he needs a convincing argument that environmental and demographic conditions have throughout history made movement in the direction of some aggregate maximizing of individual well-being a necessary condition of the continued existence of social arrangements and the societies that manifest them. Or at least he needs an argument that this has happened more often than not, or at the most crucial turns in human history. No such claim appears even vaguely plausible.

Indeed, the most convincing account of the evolution of human society that I know of proceeds on the well-supported hypothesis that this assumption is false. The evolutionary approach to anthropological theory developed in the work of Harris (1979), Cohen (1977), Harner (1977), and others is the most fully worked out functional theory of the sort Railton envisions, one quite explicit about explaining social change in terms of the interaction of individual objective interests with the changing biological environment. Harris, for example, traces an evolutionary sequence of social stages from the hunter-gatherer system, through neolithic agricultural organizations, to the pre-state village, chiefdoms, and eventually the state. Each evolutionary change represents, on his theory, an adaptation of social organization in order to meet the interests of individuals faced by environmental changes, in large part of their own making. The hunter-gatherer societies of prehistory include above 90 percent of all the human beings ever born. As Harris explains, not only was it chronologically the first social arrangement, it was the last to be characterized by what Railton would describe as optimal social rationality: "Political-economic egalitarianism is a . . . theoretically predictable structural consequence of the hunter-gatherer infrastructure" mainly because there is never any surplus to redistribute. "Egalitarianism is also firmly rooted in the openness of resources, the simplicity of the tools of production, the lack of nontransportable property, and the labile structure of the band . . . extremes of subordination and superordination are unknown. . . . Women would not be used as the rewards for male bravery [because warfare is unnecessary], sex-ratios would be in balance, and serial monogamy for both sexes would prevail." Indeed, Harris explains why "the slight degree of [human] sexual dimorphism would not impair the highly egalitarian bent of hunter-gatherer social life" (Harris, pp. 80–83).

But then environmental conditions changed in ways that would no longer support these social arrangements, so that survival of human societies required the abolition of these now maladaptive institutions that maximize well-being equally. The ice age, together with overhunting, extinguished the supply of game, and forced social groups to choose between extinction and agriculture. But agriculture increased reproductive levels and reduced mobility, thus fostering inter- and intra-group conflict. The appearance of storable and surplus food led to complex social organization, stratification, and inequities based on control of this surplus. Harris convincingly explains as adaptive for societies' survival – under these conditions – the appearance and persistence of female infanticide, obligatory marriage rules, male domination, warfare, despotic imperial state-systems, in fact, the whole panoply of past social practices we condemn as immoral, despite their enduring character. And Harris explains much of the underlying detail as well, in terms of a functional account methodologically indistinguishable from the functional theory Railton requires.

Harris's theory is not incompatible with the historical and sociological works that Railton cites. It is a much wider theory, that easily accommodates and provides a more general explanation of the sixteenth- through twentieth-century social processes which Railton's favored social scientists treat. Indeed, it may even underwrite Railton's claim that the moral goodness of social arrangements has increased over the very recent past, say, the period from the onset of the industrial revolution. For in this period, environmental and demographic forces may well have made some aggregation of well-being optimally adaptive for societies and social arrangements.

It is clear that Railton must reject this theory, for its reading of the role of social rationality in the evolution of social arrangements does not have the normative moral that Railton's does. If Harris is right, there is no reason to suppose that over the long term maximizing some aggregate individual well-being has been adaptive for society; in fact quite the opposite. But it is pretty clear that if Harris's theory is right for the whole of social history, Railton's claim, that moral goodness is whatever in general optimizes social rationality, must be false, when the latter is understood in its empirical guise as adaptation.

5   REINTERPRETING RAILTON'S MORAL THEORY

How should a defender of Railton's theory respond to this conclusion? He can either reject a theory like Harris's or revise his own account of moral goodness to accommodate it. There seems to be a way to do this latter: identify the morally good with every long-standing social institution for which an adaptation explanation is convincing; then attempt to find a way in which these institutions really did maximize some aggregate well-being of the members of their societies, no matter how repugnant these institutions may appear to us.

Rejecting Harris's view outright requires a specification of the environmental fact, which throughout human history has made aggregating well-being an evolutionary adaptation for all societies, and has made arrangements fostering inequities and disparities maladaptive. Without such a specification, Railton does not even have a challenge to Harris's theory, let alone a superior alternative. It is worth emphasizing that even if the details of Harris's reconstruction of the sequence of social institutions is wrong in detail, it is not this sequence that Railton must challenge. It is the environmental constraints, and changes in them that Harris identifies, which Railton must undercut and must find substitutes for.

Rejecting these environmental constraints is implausible. So, let us consider the ways Railton's theory might be revised or interpreted to accommodate them. Railton cannot simply say that many social institutions have persisted over long periods despite their tendency to decrease social rationality. For social rationality is supposed to be part of an explanatory theory about what has actually happened in the course of human history. A functional theory like Railton's cannot remain silent on the persistence of discontent-provoking institutions over the long term. It is not enough to say, as Railton does, that his mechanism, like those of evolution, "does not guarantee optimality or even a monotonic approach to equilibrium" (p. 194), especially in the face of theories (like Harris's) that explain departures from moral "optimality" on the basis of the same kind of evolutionary mechanism. Railton tells us, by way of excusing his theory from any commitment to monotonic improvement in the direction of social rationality, that "Human societies do not appear to have begun at or near equilibrium." Therefore, at best, "one could expect an uneven secular trend towards the inclusion of the interests of social groups." But a functional theory of the sort Railton envisions must postulate a mechanism for keeping a society within a stable equilibrium point, one determined by its environment, where this point is either itself an optimum or a sufficient condition for tracking an optimum, modulo the environment. To say that human societies do not appear to have begun near an equilibrium of any kind is incompatible with adopting a naturalistic evolutionary theory of how they change, as their environments change. The notion that, starting far from any equilibrium point, an evolving system can survive and move towards it is worse than complacent functionalism. It is a species of immanent perfectionism.

If Railton had seen the need to specify environmental contingencies against which social changes could be measured for "rationality," i.e., adaptation, he might not have made this claim. He would have seen that any evolutionary theory of biological adaptation needs to have convincing explanations of departures from apparent optimality (e.g., random drift or changes in local conditions, etc.) because it presupposes a long-term movement in the direction of optimization. To the extent such a theory lacks them, its explanatory power is minimal at best. The same must be said for Railton's theory. Thus, for example, we need to know why slavery, Railton's example of a "brutal exclusionary social arrangement" (p. 195), has persisted, not for hundreds of years, but through all of recorded history, in spite of the uneven secular trend towards social rationality.

On pain of fatally undermining the explanatory power of his theory Railton has to endorse as socially rational many of the social arrangements we hold morally repugnant today, but which typified long periods of past history, periods far longer than a millennium in some cases. Railton will have to allow that these arrangements reflected some aggregation of the objective interests of their subjects, and that therefore these apparently repugnant arrangements were, on aggregate, morally good. Now, every social system that succeeded the hunter-gatherer involved distributional inequalities, and most involved extreme social stratification, exploitation, gender discrimination, and violent preservation of the status quo. In all of these post–hunter-gatherer societies the subjective interests of the vast majority of the population certainly did not coincide with the actual arrangements, even though in most cases every one was better off – in regard to physical well-being – than had they remained hunter-gatherers (for in that case they would not have survived at all). Their objective interests thus justified the Railtonian judgment that societies characterized by slavery, warfare, and every conceivable violation of the person were morally better than they would have been had hunter-gatherer arrangements persisted. And what is more, to the extent each of these stages in the evolution of human societies was causally necessary for the ultimate appearance of still morally superior social arrangements, they were not just better than earlier states, but were morally good, *sans phrase*. Though it reconciles Railton's account with social evolution, this conclusion undermines Railton's defense against the charge of "complacent functionalism." At any rate it seriously compromises the application of the theory to criticize the most horrifying aspects of prior social practice, no matter what its power for criticizing present practice.

But in any case this tactic is not really available to Railton. He insists that moral realism must make sense of most of our moral intuitions: it should "insofar as possible capture the normative force of [our evaluative] terms by providing analyses that permit these terms to play their central

151

evaluative roles. . . . [T]hey [should] express recognizable notions of goodness and rightness" (p. 205). This is something the tactic here envisioned
surely cannot do.

Of course, Railton may want to jettison the requirement that an empirical theory of moral goodness should coincide with our relatively myopic
moral judgments for they may be no better than our often mistaken views
about our objective interests. Indeed, there is considerable impetus within
his theory for doing exactly that. After all, as Railton claims, objective interests enhance biological well-being, and well-being is what broadly conduces
to survival of the lineage of the individual. So, even though a vast majority of the members of a given society might be quite discontent with their
lot, that lot might nevertheless foster the survival of each of their individual
lines of inheritance, or at least the survival of some optimum number and
distribution of agents, given the environment, population size, available technology, etc. The result is that at every stage of human history most of the
prevailing institutions reflect the best of all possible worlds, though almost
everything in them may be a necessary evil. Expressions of discontent, whether
by organized or unorganized individuals, would reflect mistakes about objective interests and would need to be suppressed in the interests of moral
goodness.

This is true, unless, of course, at any or perhaps every stage of social evolution the objective interests of individuals are severed from their evolutionary
bases. In such a case, it may turn out that the expressions of discontent will
sometimes, perhaps even often, reflect unmet objective interests of individuals, even though they subvert social arrangements necessary for the survival
of aggregates of them or their progeny. In these cases, meeting them will be
morally good, even though it results in a heightened probability of extinction
for the whole society. But one question this raises is whether moral goodness
must take into account only agents alive at any given time and not also future generations? If the latter, we may not reconcile a theory like Railton's
with discontent that threatens the survival of the society, for satisfying such
discontent can hardly be socially rational. But if we do not take the unborn
into account in a theory about morality as social rationality, we not only deprive it of much explanatory power as an evolutionary explanation, but we
also forgo any explanation for our intuitions that the interests of our and others' unborn offspring should count for something in our moral calculations.
Surely no one supposes that moral goodness is whatever optimizes some
aggregation of the objective interests of the present members of a society, regardless of its effects on future members of the same society. Indeed, Railton

seems to include these future members when he claims that socially rational institutions optimize some aggregation of the well-being of "all *potentially* affected individuals counted equally" (p. 180, emphasis added).

6   SOCIAL RATIONALITY – MORAL GOODNESS FOR WHAT?

Individuals ought to pursue their objective interests because this is instrumentally rational for achieving their well-being, their non-moral good. Why is well-being an intrinsic end of individuals, whereas their objective interests are instrumental ends? Because nature has fostered these ends for its own evolutionary ends. Societies, too, should pursue the objective interests of the individuals who compose them, for doing so is instrumental to the society's well-being, and so, to its chances of surviving and improving in its adaptation to the environment in which it is located. Optimizing some aggregation of individual well-being is "socially rational" because it is instrumental for attaining this end of the society. So far, there is nothing distinctively moral in this sketch of Railton's theory. Morality is introduced by the identification of the socially rational with the morally good.

But this identification of the socially rational with the aggregation of individual objective interests and the morally good will be mere fiat unless two other questions can be answered in a satisfactory way. First, why is it the case that optimizing some aggregation of individual well-being is necessary for the survival of the society? Second, and more important, why should, in the moral sense of should, societies survive?

Presumably Railton's answer to the first question is simply that the well-being of societies is a function of some aggregation of the well-being of their members, and its well-being is necessary for its survival. As we have seen, this is probably false for much of human history, though it might be true for some societies at some times under some environmental conditions. But how can Railton answer the second question? This is roughly the question, what is so valuable about societies? On Railton's instrumental notion of rationality, the question can always be posed, "rational for what?" Identifying something as rational presupposes an end or objective whose attainment it fosters. Social rationality, a.k.a. moral goodness, has "reality" on this theory because a society's attaining or increasing it is an end instrumental to society's survival. But is the survival of society an instrumental end or an intrinsic one? If the former, then there is some other end, either of society or of some larger system within which it figures that must be identified.

And if it is an intrinsic end, then all the problems that haunt traditional conceptions of moral realism return to unravel the whole of Railton's new argument for it.

To show that the survival of society is instrumentally rational for the biosphere, or ecosystem, or some other system that includes it, requires showing that its survival is necessary for the optimization of the well-being of the entire ecosystem. Even if, in the face of nuclear winter, overpopulation, the greenhouse effect, chemical pollution, destruction of species, etc., we could show this, the question of intrinsic vs. instrumental ends must recur at the level of the ecosystem as a whole. At some point, some end or other will have to be identified as morally good in itself. So, we might as well face, sooner rather than later in the regress, the question whether anything can be good in itself. This is the question arguments for moral realism must answer, for no one ever doubted that there are things that are good as means.

But Railton's argument for moral realism is anchored on showing the causal role of moral goodness in a functional network – one in which moral goodness is a means, and not an end. Even if the theory were empirically convincing, it would not succeed in showing that what he calls moral goodness is morally good.

In a way Railton recognizes this fact. He recognizes that whether social rationality is really morally good is, in Moore's phrase, an open question: Such "open questions" cannot by their nature be closed, since definitions are not subject to proof or disproof. But open questions can be more or less disturbing, for although definitional proposals cannot be demonstrated – they can fare better or worse at meeting various desiderata (p. 204). The desiderata Railton believes his definitional identification meets are actually adequacy conditions on the social theory in which the definition figures. But the theory does not, I think, meet them. First note that the issue of whether the identification is a definition or a central theoretical claim in a social theory with positive and normative ramifications is, of course, irrelevant to whether the theory meets adequacy criteria. To describe the identification of social rationality with moral goodness as a definition seems to ward off the question, "Is social rationality really everywhere and always morally good?" as silly, rather like the question "Is social rationality really social rationality?" But this ploy will not work among naturalists. For if there is one thing they agree on, it is that when it comes to scientific theories, the distinction between those sentences which are treated as definitions and those treated as axioms of the theory comes to very little. So the real test of whether this "definition" is a good one, or equivalently, whether this "axiom" is a well-justified one, is the explanatory and predictive power of the theory in which it figures. This is a point with which Railton, of course, agrees: "I have assumed throughout that the drawing up of definitions is part of theory-construction" (p. 204). So, in Railton's conclusion, whether definitional or not, his proposal is said to meet the following criteria on theory construction in this domain (the labels on these criteria are mine):

(1) Explanatory constraints: (a) Does the theory in which the definition figures systematize a large enough proportion of our actual judgments of moral goodness and rightness? (b) Do the moral notions defined in the theory actually do any explanatory work in systematizing the theory's subject matter?

(2) Evidential constraints: "[Are] the empirical theories constructed with the help of these definitions ... reasonably good theories, that is theories for which we have substantial evidence and which provide plausible explanation?" (p. 205).

If my account of the evidence is closer to correct than Railton's, then neither Railton's actual theory nor a reasonable revision of it can jointly meet these two sets of criteria. The best theory I know of consistent with the evidence identifies as instrumentally rational for the survival of societies many institutions we could never approve as morally good nor explain as persisting because of their moral goodness.

This leaves us with the question of whether the defects here alleged are restricted to Railton's particular attempt to yoke together empirical social theory and moral philosophy. Or do they reflect obstacles to any attempt to substantiate a naturalistic moral theory by means of social science? I venture to suggest that the moral to draw from this essay is not restricted to Railton's approach. The reasons are twofold. First, to be at all attractive to us, any moral theory will have to reflect our most basic moral beliefs. Secondly, the only tractable naturalistic theories in social science, and psychology for that matter, are all functional ones, of roughly the sort Railton appeals to. Since they are predicated on the assumption that their objects, individual people or whole societies, are either in a stable equilibrium that optimizes some quality or are moving towards such an equilibrium, the moral realist has little choice but to link goodness with this optimized quality. To do otherwise deprives moral goodness of its explanatory role in the empirical theory, and so deprives it of its claim to reality. This means that every attempt to harness a theory of the good to an empirical social theory must face the charge of complacent functionalism.

NOTES

Reprinted with permission from *Midwest Studies in Philosophy*, vol. 15, edited by Peter A. French, Theodore E. Uehling, Jr., and Howard K. Wettstein (Notre Dame, IN: University of Notre Dame Press, 1990), pp. 150–166.

1. This account of ethical naturalism is to be found, for example, in Nicholas Sturgeon, 1985, pp. 49–78, and Richard Boyd, 1988.
2. Boyd's "How to Be a Moral Realist" provides an example both of this view about the relevance of social science and the one outlined in the previous paragraph about the importance of naturalism to realism: "according to any naturalistic conception [of morals] . . . goodness is an ordinary natural property, and it would be odd indeed if observations didn't play the same role in the study of this property that they play in the study of all the others. . . . [G]oodness is a property quite similar to the other properties studied by psychologists, historians, and social scientists . . ." (Boyd, 1988, p. 206).
3. Boyd is clearly committed to the same type of functional approach in his account both of individual and social goodness, although his "homeostatic" theory is not worked out in the sort of way that will reveal its connections to current social theories nor to the methodological problems that such theories face. See Boyd, 1988.

8

Contractarianism and the "Trolley" Problem

The "trolley" problem was introduced by Judith Jarvis Thomson in two papers, "Killing, Letting Die and the Trolley Problem," and "The Trolley Problem" [Thomson, 1986; see also Foote, 1967]. The puzzle raised by the "trolley" problem is that of finding a principle that will reconcile two strongly held intuitions: that in the case of a trolley hurtling down the tracks towards five innocent persons, it is permissible to shunt the trolley onto a siding, thereby killing one person so that the five may live; while in the case of five patients requiring body parts, it is impermissible to anesthetize an innocent person and transplant the needed organs.

The problem is serious because the moral principle that most obviously seems to underlie the "trolley" case, viz., it is permissible to kill one in order to save five, is the direct denial of the principle that seems to underlie the "transplant" case. Indeed, the two cases are often supposed to place sharply in contrast consequentialist and deontological moral principles. For, in "trolley" the greater good for the greater number seems to decide the case, whereas in "transplant," the rights of the innocent trump the welfare of the patients. Like Thomson, other philosophers have sought a principle that will reconcile our intuitions about "trolley" and "transplant." Optimally, such a principle should *explain* why there is almost universal agreement on the two cases – that killing the one to save the five is permissible in "trolley" and impermissible in "transplant" and *justify* the moral distinction that we draw between them. It may of course turn out that there is no sustainable moral distinction between these two cases; that either there is no difference because both actions they describe are permissible or both impermissible, or that there is no moral fact of the matter to justify. If this were so, the justificatory problem would disappear, but the explanatory problem would still be left to us. For it is widely reported among philosophers and others who put this problem to audiences, that there is near universal agreement about the two thought experiments. And this is a fact of the matter that needs explanation.

In "The Trolley Problem" Thomson suggests that the morally relevant difference between "transplant" and "trolley" is that (i) in "trolley," the bystander is diverting a preexisting threat, while in "transplant" the physician is creating a new one, and (ii) in diverting the threat the bystander does not use "means which themselves constitute an infringement of any right of the individual who dies as a result of the diverted threat." Presumably, this suggestion can be built up into a full explanation and justification of our intuitions in these two cases. There are at least three problems with this suggestion: (a) there may be counterexamples to it, which I shall describe below; (b) it is unexplained what the moral difference is between diverting a preexisting threat and creating a new one, (c) there will be disagreements among interlocutors about whether the victim in "trolley" has rights that are being infringed, and these disagreements will undermine the explanatory adequacy of Thomson's suggestion or its development. In what follows I attempt to sketch a more fundamental explanation and justification for our intuitions in "trolley" and "transplant," to show how it circumvents the problems facing Thomson's suggestion, and to suggest how my proposal may accommodate her own. My proposal is a contractarian solution to the "trolley" problem. A conflict between consequentialist and deontological theories is an apt invitation to contractarian approaches. In fact, given the strength with which people embrace the intuitions of "trolley" and "transplant," together with the evident difficulty of either consequentialism or deontological theories adequately to explain and underwrite these intuitions, the ability of a contractarian theory to do so would be an especially telling argument in its favor as a general foundation for morality.

A contractarian solution to the "trolley" problem is one which shows that a rational individual would bargain for social institutions that did not minimize the risks of being a "trolley" victim, but did minimize the risks of being a "transplant" victim. One way to minimize a risk like "trolley" or "transplant" is to make them contractually impermissible; another is to build into agents a strong repugnance to engaging in them or approving their outcomes – one reflected in widely shared intuitions. If we can show that a rational agent would bargain to reduce the risk of "transplant" to its minimum, but would not do so for "trolley," then we have discovered a morally significant difference between them. Of course, it is a long step from this difference to an explanation-*cum*-justification of our moral intuitions about "trolley" and "transplant." I explore this idea at the end of this chapter.

Of course, the institutions minimizing the risk of "transplant," while leaving the risks of "trolley" alone to be chosen by the rational agent, must be described far more generally. For the conflict here is not between intuitions about trolleys and transplants, or even transportation and health; the details are merely illustrative, and rational agents bargaining to establish a commonwealth cannot be supposed to know anything beyond the most general facts about themselves and their world. In what follows I will assume that agents understand the basic causal structure of our world. In fact, this knowledge is crucial to their deliberations. Nothing in what follows should turn on assumptions about the degree to which rational agents are risk-averse, so long as they do not positively relish putting their lives at risk.

What a contractarian approach must do is justify as most preferable to the rational agent a package of institutional rules that allows the agent to be the innocent victim in some kinds of cases and not in others, where the difference between these cases sorts "trolley" and "transplant," and all the other cases like them (and between them), in a way that corresponds to our intuitions. But the difference between the cases need not be a moral difference. Indeed, if the solution to our problem is really contractarian, it will not be a moral difference, but some prudential one, reflecting factual differences between the cases.

1 COSTS, BENEFITS AND FREQUENCIES

Let's begin with the special case of runaway trolleys and unauthorized transplants. Then we can turn to the more general cases they instantiate. The contractarian seeks a difference between "trolley" and "transplant," in the state of nature so to speak, that makes it reasonable to contract for social rules which reduce the risk of being the innocent victim in the "transplant," but not in "trolleys." These differences are not difficult to find. Consider first the costs and benefits of exposure to the risks in each case, and the costs and benefits of complete protection against risk of death in either case. We calculate costs and benefits by determining net expected utilities – the products of utilities and disutilities with their respective probabilities.

In "trolley" the cost of exposure is the value of one's life times the very low probability that one will actually be a victim of a runaway "trolley" shunted onto a track one happens to be crossing. The benefit of exposure is the value in convenience of crossing trolley tracks times $(1 -$ the probability of being fatally struck by a shunted trolley), a number very close to one. It is clear that the net benefits of exposure outweigh the net benefits of avoiding all risk of exposure to trolleys. Anyone who demurs from this conclusion is condemned by consistency of reasoning in other cases to being a total recluse, never venturing forth: whenever one leaves one's bed, one is almost always at some infinitesimal "trolley"-like risk.

Consider "transplant": the cost of exposure is the value of one's life times the probability that one will be unwillingly deprived of one's organs. The benefits of exposure are those which accrue to health care by a physician times (1 – the probability of being unwillingly deprived of one's organs). The costs of complete protection against the deprivation of one's organs will be the reduction in health consequent on not consulting physicians times the probability that avoiding physicians will prevent the alienation of one's organs, a number equal to one. Here the calculation is much harder to make. After all, avoiding medical attention will secure one against unauthorized organ deprivation, but also fate one to the longevity of those who believe in faith healing. If the misappropriation of organs is a very frequent occurrence, the cost-benefit calculation will lead rational agents to trust their fates to faith healers.

So, the rational agent needs to determine the utilities and disutilities associated with the outcomes of "trolley" and "transplant," and the probabilities associated with them, in order to calculate an expected value. If it is positive for either, he will accept institutional rules that make them permissible, and if it is negative for one or the other, he will favor rules prohibiting the negative outcomes, in order to make the net expected value of using trolleys or physicians come out positive. Assigning utilities and disutilites in these cases is of course difficult. The disutility of being killed by a runaway trolley is about the same as the disutility of being killed by a surgeon seeking organs for transplant. The utility derived from consulting a physician seems to be higher than the utility derived from skulking around trolley sidings. But these latter examples are not the relevant utilities to consider. We must compare outcomes of equal descriptive specificity in assigning utilities, because almost any outcome comes under many descriptions, some of which will be permissible and others impermissible or even obligatory. So, the relevant comparison is between the utility of being able to cross trolley tracks versus the utility of consulting transplant surgeons. In a world anything like ours it is practically impossible to use a trolley system without crossing a track, so the utility of being able to do so may be very high indeed. By contrast, in a world like ours the utility of being able to consult a transplant surgeon is either quite low or indeterminate. I for one would sooner give up the opportunity to consult transplant surgeons than I would give up the opportunity to cross trolley tracks. Consider a more general description of each alternative. With which is more utility associated: the opportunity to use conveyances, or the ability to consult physicians? Here the choice seems clear. For a life on foot is the only alternative to transport, while if we cannot consult physicians, we can take steps to protect our health in other ways. If you disagree with my ranking, consider how many more people who can otherwise afford modern medicine forgo it, compared with the number of people who willingly shut themselves in to avoid the hazards of life out of doors. In any case, for present purposes it is enough to assume that the utility of using trolleys is about the same as the utility of consulting transplant surgeons. For the real difference between "trolley" and "transplant" is to be found in the probabilities.

The frequency with which current levels of medical care exposes us to misappropriation of our organs, or to similar insults, is vastly higher than the frequency with which skulking on tracks exposes us to runaway trolleys shunted to save the life of one or more persons. By and large our knowledge of medical practice is limited. Many of us cannot be sure whether the blood drawn from us under pretext of a Wassermann test might not be used for someone else's transfusion, or to culture a line of cells that will enrich the physician. When we put ourselves in the hands of a physician, we allow ourselves to be used as means to the physician's ends. The result is only very rarely injurious to us, but the fact is almost every time we undergo some extensive medical treatment there is some chance that our bodies or their parts could be diverted to serve the interests of others. As medical science improves this is a risk that will increase. Before the advent of immune-reaction suppressants, there was no trade in body parts at all. Once immunity problems are completely solved, the potential for misappropriation of organs will increase greatly, and so will the frequency at which we put ourselves at risk of such misappropriation. Even the knowledge that one has five times the a priori chance of being the recipient of stolen organs than the victim is not enough to make it rational to accept such risks. For the frequency with which each of us puts ourselves at risk is so great – we go to the doctor so much more often than we need new organs, that the prospect of securing them by unwilling transfer is not enough to compensate for the risk of being the victim of such transfer. Of course, if the only reason one ever sought medical attention were for the transplantation of organs, then it might make prudential sense in general to allow unconsented transfer of organs from one unwilling person to two or more patients. But since we see doctors for many reasons and with considerable frequency, and since whenever we do so, our organs could be misappropriated, the probability of receiving organs from unwilling donors times the utility of doing so cannot be as great as the probability of losing vital organs unwillingly times the disutility of doing so.

If the frequency of putting oneself at risk of "transplant" is high enough, one will forgo the tender mercies of medical treatment. On the other hand, the rational agent will seek to reduce those risks to acceptable levels, because of the benefits that attend medical consultation.

So, here are relevant differences between "trolley" and "transplant" in the state of nature: it would be inconvenient to avoid all "trolley"-risks, but it would be fatal to avoid all "transplant" risks. "Trolley"-risks are extremely improbable. "Transplant" risks may be improbable, but they are much more probable than "trolley" risks. In the improbable "trolley" scenario, each of us has a higher a priori chance of being among the five saved than the one killed in a "trolley" situation. In "transplant" each of us has a higher probability of being victimized at least once in the course of many visits to the doctor than we do of being on the siding of "trolley," and we have no clear idea of what the probability is that we will require an organ transplant some time in our lives when the only source is an unwilling donor.

Given these costs, benefits, and probabilities, the rational agent will seek means of reducing chances of "transplant" to the minimum, while already recognizing that "trolley's" chances are low and favorable to him in any case. And one way to do this is for the rational agent to bargain for rules that will make "trolley" permissible and "transplant" impermissible.

2 THE GENERAL CASE

We are of course not really interested in trolleys and transplants, but in more general issues. The rational agent contemplating social arrangements he or she would find attractive will not even focus on such general problems as public transport and major surgery. Indeed, the moral principles that make "trolley" permissible and "transplant" impermissible can't hinge on any detail about so general a matter as the fact that agents will want to move about or will need health care. For it is easy to construct "trolley"- and "transplant"-type cases that have nothing to do with locomotion and biology.

So, what are the features of "trolley" and "transplant" that are shared in common with every case like them? They are facts about causal order, about being harmed as a cause or an effect of events that help others. In "trolley" there is the chance of dying *after* and as an unavoidable *result* of actions that save another. Notice that this chance is a different one from the chance that one has of one's death *bringing about* that another person is saved. In our actual world, given social and causal arrangements of the sort we are familiar with, there are far more ways of saving *n* people *by* actions that kill

I different people, than there are ways of killing I people *by* actions that save n different people, where $n > I$. (In what follows I abbreviate this point by saying that saving n people by killing I people is a much more frequent occurrence than killing I people by saving n people. But to be accurate the contrast is not between killing by saving and saving by killing, but between killing by actions that lead to saving of lives, and saving by actions that lead to killing. I am indebted here to Judith Jarvis Thomson.) It is this latter risk that "trolley" involves, and the former risk that "transplant" instances. This asymmetry in frequencies reflects a relatively deep "metaphysical" fact about dispersal of causal order in a world like ours that obeys the second law of thermodynamics. A world in which saving n lives unavoidably leads to I deaths, with a long-run frequency much above zero, would look to us like one in which a lot of Rube Goldberg devices mysteriously came into play with uncanny frequency, and without any cause systematical enough for people to do much about (for certainly people would try to do something about this type of causal sequence). But since we can't reduce this frequency, this is a world in which a certain pattern or orderliness (the saving of n lives causing the losing of I lives) recurs with a frequency that violates the second law of thermodynamics' requirement of overall increase in disorder, in entropy. By contrast, for almost any accidental death we can think of we can also imagine physically possible non–Rube Goldberg arrangements whereby the death could have been prevented by the immediately prior sacrifice of another life. This difference in our beliefs about the frequency of causal sequences is reflected in the fact that we have instituted posthumous awards for heroism to those who have sacrificed themselves for others, but no special insurance indemnity for those who die accidentally as the result of actions which save the lives of others.

The division of labor, the return to scale of people working together, the productivity associated with using powerful and potentially dangerous devices with due care, provide substantial utility for large numbers of individuals at extremely low risk to any of them of being killed as a result of actions that save a larger number of people. Note in most of these circumstances of modern life, the only way to protect all individuals equally from the risk of being killed as a result of actions that save others is by depriving all individuals of the benefits of these conveniences. So, presented with the choice between (a) such institutions as the division of labor, returns to scale, technology, plus the risk of being made a victim of such technology as a result of the saving of the lives of others, and (b) no risk and no such benefit, the rational agent will adopt such institutions, provided the risk is low enough, as indeed it is in the actual world.

In short, each rational agent who is party to a contract for establishing social institutions will accept rules that render permissible the sacrifice of any party to the contract as a result of the saving of five other parties to it, provided benefits associated with this risk are high for everyone, the frequency with which any one party is put at this risk is extremely low, and the costs to everyone of lowering this risk to zero for any party are prohibitively high for everyone. This is the consideration about the general case that eventuates in the permissibility of "trolley" and many other cases like it which have nothing to do with public transport, locomotion, or any other particular feature of a complex society. When social arrangements can put us at the risk of others using us as a means, rational agents will want strong protections against doing so, especially, when (as with health care) the costs of not putting ourselves at such mercies are great. The rational agent who is party to a contract will choose institutional arrangements that prohibit the sacrifice of one party to the contract to save some other party or parties to it, provided that the frequency with which each party to the contract is put to such a risk of being used is high, and the costs to everyone of lowering this risk for him or herself (by avoiding all exposure to it) are also extremely high.

In short the difference between "trolley" and "transport" is that the calculation of expected value conducted by the rational agent in choosing a social contract for our world will result in the selection of rules that make "trolley" permissible and "transplant" impermissible.

It is worth repeating that the rational agent will not need to know very much about the state of nature in our world to see that the frequency of "transplant"-type risks – where the death of one causes more than one to be saved – will be greater than the frequency of "trolley"-type risks. The differences will be apparent long before anyone dreams of the possibility of trolleys and transplants. If I am right that these are the relevant differences, then in a possible world so different from our own that the costs, benefits and frequencies of "trolley" and "transplant" converge and change places, the rational agent should prefer arrangements that make "transplant" permissible and "trolley" impermissible, because the mathematical expected values associated with the risks of "trolley" and "transplant" come out differently. That is, if the rational agent's prospects of being killed on "trolley" track sidings as a collateral effect of saving the lives of others is quite high, then, other things being equal, the less willing should the rational agent be to agree to a rule which makes victimizing people on sidings permissible. The more frequently everyone expects to be in need of an unwilling donor's body part, and the less frequently everyone expects otherwise to need to put oneself at the mercy of a physician, the more tempted one will be to accept institutions that make "transplant" permissible. Similarly, as the net payoffs to exposure to the risks of "trolley" and "transplant" change, the rational agent will be willing to countenance other rules, rules which make "transplant" permissible and "trolley" impermissible. Thus suppose most agents live a life of drudgery and boredom and suicide is frequent. Suppose some very large number require organ transplants, and suppose further that transplanted organs have a wonderful effect on people, making their lives worthwhile and meaningful, but they only have this effect if they come from unwilling donors. Then rational agents, knowing they were to find themselves in such a world, would choose arrangements which made "transplant" permissible. And if in this world, people were frequently and unavoidably put at risk of dying as the result of almost any action that would save the lives of others, rational agents would seek arrangements that prohibited interference with the course of nature, no matter how benevolent the intention, no matter how ignorant the Samaritan was of its inevitable consequences.

What seems reasonable then is that as *a priori* estimates of probabilities and utilities change, social rules that prohibit "trolley" will recommend themselves to the parties to the contract, and ones that permit "transplant" will do so as well. If this is right, it is a powerful consideration in favor of the contractarian approach.

It can also explain why diverting a preexisting threat versus creating a new one *seems* to be a morally relevant distinction in our world, but would not be in a world differently arranged. Threat-creation is impermissible in our world largely because the opportunity to create a new threat to one that might save more than one arises far more frequently than the opportunity to divert a preexisting threat from five to one. Therefore, we are all more frequently at risk of newly created threats than diverted ones, and *ceteris paribus* the rational among us will prefer institutions that protect us from the more frequent threat. One way a rational agent will do this is by agreeing to assign rights in a way that affords protection against the more serious threat. Whence Thomson's approach to the "trolley" problem, which makes diverted threats permissible and newly created ones impermissible.

3   OTHER CASES

Still another way to test this solution to the "trolley" problem is to expose it to the gamut of cases that have been constructed since "trolley" and "transplant" were first brought forward. "Drug" is a case in which one must give a single saving dose of a medicine either to one person, or divide it among five, who will be saved by the smaller doses. It is certainly permissible to give the drug to the five and let the one die. "Fatman" is a case in which the bystander may stop the trolley by pushing a large person onto the track, instead of diverting the trolley onto a siding where it will run over a solitary person. Here it is evidently impermissible to push the person onto the train.

I suggest that the reason we rule these cases the way we do is to be explained by how a rational agent would decide on social rules, given the probabilities of finding ourselves under the descriptions relevant to these cases and given the utilities and disutilities associated with exposing ourselves to these cases.

The relevant rule in "fatman" is not one about what the rational agent would sanction be done to him or her if he or she were fat and in the right place at the right time to stop a runaway trolley headed for five people. Rather, the specific consideration which a rational agent endorses that makes "fatman" impermissible is the same principle that prohibits "transplant": the *frequency* with which the use of public transport puts one at risk of being used somehow or other to save two or more people at considerable cost to oneself is *far greater* than the frequency with which two or more persons being saved from death by public transport puts a third person at risk of death. This is just another example of the general causal fact about our world that enables the rational agent to decide which institutions to endorse: in our world the frequency of cases of saving more than one by causing cases of dying by one is far lower than the frequency of dyings by one being caused by savings of more than one.

John Martin Fischer offers the following objection to Thomson's solution to the "trolley" problem: Consider "ramp" – a case in which one can save the five by pushing a button which activates a ramp that sends the trolley upwards on a collision course with a bridge over the tracks, on which stands the fatman, from the case of that name. According to Fischer, if shunting the trolley onto the siding is permissible, then so is "ramp" – after all one moves the trolley to the side, and the other moves it up and out of harm's way, and there can be no morally relevant difference between upwards and sideways. On the other hand, "fatman" – saving the five by shoving the man onto the train – seems impermissible, because it violates a stringent right. In this it is like "transplant." But, argues Fischer, "ramp" and "fatman" differ only in morally irrelevant ways – in "fatman" the man is shoved on the trolley, while in "ramp," the trolley is shoved on the man. Though Fischer does not appeal to the special theory of relativity, it is pretty obvious that there is no morally relevant difference between treating the man at rest and the trolley in motion, and vice versa. Thus, any difference between "fatman" and "ramp" is relative to a morally irrelevant reference frame. Fischer infers that creating a threat and violating a stringent right cannot be what makes "fatman" impermissible, since apparently no such right is violated nor any threat created in "ramp," and none could be so long as we hold "ramp" indistinguishable from "trolley." Accordingly, there is a slippery slope, claims Fischer, all the way from "trolley" to "transplant," which shows that creation versus diversion of threats and violation versus respect for stringent rights will not explain or justify our intuitions [Fischer, 1991].

But, given the relevant frequencies of event-types under which these cases fall in our world, a rational agent contemplating the acceptance of the risks these event-types raise, will be able to identify differences between "ramp" and "fatman" that stop the slide down Fischer's slope. Assume that the utilities and disutilities of exposure to "ramp" and to "fatman" are about the same. Then the question of whether one or both or neither is permissible will hinge on frequencies and differences in them. In our world, "fatman" is an instance of a far more common occurrence than "ramp." For it will be more frequently the case that someone's death is or could be causally prior to saving several lives than it will be the case that the saving of several lives is or would be causally prior to someone's death. Thus, expected value to the agent of rules permitting "ramp" is greater than ones permitting "fatman." If in fact the difference between them is that the expected value of ramp-permitting rules is positive and of "fatman"-permitting rules is negative, then the rational agent's contractarian choice will obstruct Fischer's slippery slope. There is after all a prudentially significant difference between "ramp" and "fatman." And for a contractarian, that's enough of a difference to be a morally relevant one.

4   CONTRACTARIANISM AND DOUBLE EFFECT

It's worth pointing out the differences between the contractarian decision about principles to be endorsed for the arrangement of social institutions and the doctrine of double effect. This principle, it widely has been agreed, will not solve the problem of adequately squaring our "trolley" and "transplant" intuitions. The doctrine of double effect supposedly underwrites "trolley" because the death of the one individual on the siding was not intended, even if foreseen as an effect of saving the lives of the five, whereas the death of the unwilling organ donor in "transplant" is intended.

One difference between the doctrine of double effect and a contractarian approach is that the former focuses on cause and effect relations, whereas the latter involves costs and benefits as well. Moreover, the relevant fact which makes "trolley" permissible and "transplant" impermissible on the doctrine of double effect is not that the death of the one in "trolley" is an effect, and the death of the one in "transplant" is a cause – both are effects – but that the latter is intended and the former isn't. For contractarianism the crucial issue is not intent, but the direction of causation. And the direction of causation is relevant because of the difference in frequencies with which savings cause deaths and deaths cause savings. There is however a possible relation between the doctrine of double effect and the frequency issue, one which may explain the appeal of double effect as a moral principle. One reason intention arises in the impermissible cases is that the frequency of deaths causing savings is high enough that people come to realize that the causal chain provides a practically useful means of saving lives, and this realization is a necessary condition for the formation of an intent. By contrast, we recognize that the frequency of savings causing deaths is so low that it provides a highly unreliable means of doing away with someone. Accordingly, we think it unlikely that when life-saving causes a death that intent was involved. This unlikeliness makes us comfortable with the rare cases in which death is foreseen, but which we can assimilate to unintended and therefore excused killings. Whence the attractions of a doctrine of double effect. It does mirror a part of the considerations that sort the cases correctly.

There is another important difference between double effect and a contractarian approach, one that has to do with costs and benefits. To see this consider Thomson's objection to the doctrine of double effect as a solution to the "trolley" problem. It turns on a counterexample, "hospital": five patients' lives could be saved by the manufacture of a drug in the hospital, but the manufacture would release lethal fumes into the room of another patient. Evidently the doctrine of double effect permits "hospital," since the release of the gas is unintended. But "hospital" does not differ morally from "transplant," and the right principle that disallows "transplant" must disallow "hospital" as well. Our contractarian will choose principles that rule the same way on "transplant" and "hospital," and differently on "trolley." Here the relevant difference is not one of frequencies but of costs and benefits. Presumably the frequencies of "hospital" sequences and "trolley" sequences are comparable. They are both cases where savings cause deaths. So the difference can't be a matter of frequencies. But there is a cost/benefit difference between health care and transportation that is relevant, and this difference in net benefits will swamp the differences in probability when the rational agent comes to calculate expected utilities. Our need for health care is less frequent but more pressing than our need for public transport, and there are no alternatives to it. The costs of forgoing health care are very high, and the benefits of accepting it are equally high. So, the rational agent will seek to reduce risks associated with making use of it. Among them will be a statutory prohibition against harms which might befall us as a result of making use of health care, and this includes exposure to lethal gases.

5   EXPLAINING OUR INTUITIONS

Given some basic facts about frequencies, costs, and benefits in our world, the contracting parties will choose those arrangements which are reflected in our own intuitions about "trolley," "transplant," and the others. How will this contractarian justification of our intuitions explain them? Here I can only sketch a speculative evolutionary scenario. *Homo sapiens* have been selected for fitness, and the rational pursuit of self-interest – of the sort contractarians appeal to – enhances fitness to a considerable degree. Accordingly, in many contexts natural selection will favor those behavioral strategies that reflect self-interest, regardless of whether these strategies reflect actual occurrent or conscious calculation of costs and benefits on the part of agents. Of course, those groups will survive whose institutions reconcile the self-interested behavior of members in their interaction with one another. Thus, interaction-rules and institutions will be favored which reflect individual self-interest, and of course by a feedback process in the long run, individuals will be selected as they order their interactions in accordance with the rules and principles governing such institutions. (In fact, if the rational agent's preference function includes as a highly preferred alternative the welfare of immediate genetic offspring, as both evolutionary biology and casual observation suggest it does, then the emergence of at least some other-regarding behavior is just what evolutionary theory would lead us to expect. This behavior will reflect the evolutionary selection of moral principles on the basis of their individual adaptational value.) Our emotions and intuitions are themselves presumed by this explanatory argument to be selected for to the extent that they reinforce and reflect the acceptance of such institutional rules. Whence our intuitions about the "trolley" problem.

I have no illusions that this is a very well confirmed or powerful and detailed explanation of the emergence of our intuitions. Like democracy, though it may not be good, it is better than any of its competitors. (I discuss the strengths and weaknesses, mainly the latter, of such evolutionary explanations-*cum-justifications* of moral principles in Chapter 8, "The Biological Basis of Ethics: A Best-Case Scenario.") No matter how good an explanation of course, an adaptational story about why our intuitions make "trolley" permissible and "transplant" impermissible is not a justification of these intuitions. In fact, the argument of this paper moves in exactly the opposite direction: the fact that contractarianism seems to explain these intuitions is at least part of a justification of contractarianism as a moral philosophy.

6 APPENDIX: SOME NUMBERS

To see how the contractarian's calculations would come out in a way that renders "trolley" permissible and "transplant" impermissible, consider the following assignment of notional utilities and probabilities: Assign the opportunity to use trolley transport the value 1, and assign to death by being run over on trolley-siding the value 1000. Assign to such a death the probability of one in a million. Thus the probability of not dying while making use of trolley transport is .999999. Then the expected value of using trolleys is .999999 less the expected value being killed while doing so, .001, which is a positive net value of .998999.

Assign the opportunity to consult a surgeon the value 1, and assign to death at the surgeon's hands disutility of 1000. Now, it is reasonable to assume that in "the state of nature" the chances of being the victim of unscrupulous physicians is far greater than the chance of being run over by a runaway trolley. Suppose it's one thousand times greater. That is, suppose the probability of being cut up for organs is one in a thousand instead of one in a million. Then the expected value of safely consulting a surgeon is 1 times .999, i.e., .999 and the expected value of dying at his hands is 1000 times .001, or 1. The net expected value of "transplant" is therefore negative, while the net expected value of "trolley" is positive. Accordingly, the rational agent will not consult surgeons unless he can change the odds or the utilities. On the other hand, the agent will use trolleys. One way to lower the odds against being cut up for the benefit of others is to see to it that all people have an abhorrence of such action.

Of course if the utilities and disutilities are different, the calculation comes out differently. In this world the chance of dying as a result of actions that save the life of others does seem to be at least a thousand times less frequent than the chance of living as the result of actions that lead to others' deaths (whence posthumous medals for heroism). Since "trolley" is an instance of the former sequence and "transplant" an instance of the latter, the probabilities are as required. If the utility associated with consulting surgeons is higher than the utility of using trolleys, then of course the calculation may result in a positive value for "transplant" after all. But consider, which would you give up? The opportunity to go abroad and not be shut in, or the opportunity to consult physicians? Most people would choose the first. After all, the life of a shut-in is hardly worth living, even if one has the constant attendance of physicians; while a person who knows that physicians will be proscribed to him or her can take precautions, for instance, train oneself in medicine, and take other steps to minimize risks. So, even if the odds between "trolley" and "transplant" are less than a thousand to one, the utility differences can swamp them.

NOTE

Reprinted by permission of the *Journal of Social Philosophy* from vol. 23:3 (1992): 88–104.

9

Does Evolutionary Theory Give Comfort or Inspiration to Economics?

To some economists, evolutionary theory looks like a tempting cure for what ails their subject. To others, it looks like part of a powerful defense of the status quo in economic theory. I think that Darwinian theory is a remarkably inappropriate model, metaphor, inspiration, or theoretical framework for economic theory. The theory of natural selection shares few of its strengths and most of its weaknesses with neoclassical theory, and provides no help in any attempt to frame more powerful alternatives to that theory. In this chapter, I explain why this is so.

I begin with a sketch of the theory of natural selection, some of its strengths and some of its weaknesses. Then I consider how the theory might be supposed to play a role in the improvement of our understanding of economic processes. I conclude with a brief illustration of the problems of instantiating a theory from one domain in another quite different one, employing the most extensive of attempts to develop an evolutionary theory in economics. My pessimistic conclusions reflect a concern shared with economists who have sought comfort or inspiration from biological theory. The concern is to vindicate received theory or to underwrite new theory against a reasonable standard of predictive success. Few of these economists have noticed what the opponents of such a standard for economic theory have seen, that evolutionary theory is itself bereft of strong predictive power (see McCloskey 1985, p. 15).

Two things to note and set aside at the outset are the historical influence that economic science has had over evolutionary theory from before Darwin to the present day, and the profit that biologists have taken in recent years from developments in economic theory. The influence of Adam Smith and Thomas Malthus on Darwin is well documented. Indeed, this is so well established that more than one opponent of evolutionary theory has attempted to tar it with the brush of laissez-faire capitalism (see Rifkin, 1984). (This was, of course,

172

before the fall of the socialist economies of the east.) Darwin himself relates the influence of Malthus on his notion of the survival of the fittest. In recent years, biologists have exploited some significant mathematical results relating to conditions of stability for general equilibria and have adapted insights from game theory to identify evolutionarily stable strategies in animal behavior. Everything I shall say hereafter is perfectly compatible with economic thought having a significant impact on the improvement of biological theory. It is just that the terms of trade are always in the direction from economics to biology and not vice versa. Why is that?

To answer, we need first a brief introduction to the theory of natural selection. This should not be difficult. The theory is breathtaking in its simplicity. It is so easy to understand that Darwin certainly need not have traveled for five years around Latin American to have hit upon it; an hour on the Sussex Downs would have sufficed. Indeed, when Thomas Huxley, one of the theory's most vigorous early exponents, first heard its details, he complained, "How stupid of me not to have thought of it."

1   NATURAL SELECTION

Darwin began with some observations. The first is the Malthusian point that organisms reproduce geometrically, and yet the population of most species remains constant over time. From this it follows that there is a struggle for survival, both within species and between members of differing species. Darwin's second observation was that species are characterized by variation among the properties of their members. Darwin inferred that the survivors in this struggle are those variants most fitted to their environments – most able to defend themselves against predators, find shelter against the elements, provide themselves with food, and therefore most able to reproduce in higher numbers. If these traits are hereditary then they will be represented in higher proportions in each generation until they become ubiquitous throughout the species. This will be especially true for hereditary traits that enhance an organism's ability to procure mates and otherwise ensure the reproduction of fertile offspring. It is crucial to Darwin's theory that variation is large and blind – in any generation there will be differences on which selection can operate, and these differences are not elicited by environmental needs, but are randomly generated. Darwin knew nothing of genetics, and his theory requires only that there be variation and heredity. Modern genetic theory provides for both of these requisites of Darwin's theory. For this reason, genetics is often treated as part of evolutionary theory. Selection, for Darwin, is a misleading metaphor, for his theory deprives nature of any purpose, teleology, design, or intentions of the sort the notion of selection suggests. Nature selects only in the sense that the match or mismatch between the environment and the fortuitously generated variants determine survival and thus reproduction.

So much for the bare bones of the theory. Now some of the details. Darwin knew little about the sources of variation. We know more. Some variation is produced by mutations, but not enough to account for the diversity and the adaptations we actually observe. Most variation, especially among sexually reproducing organisms, is the result of the shuffling of genes through interbreeding across the species. The interaction of different genes with one another and with various features of an environment produces a range of phenotypes that are selected for in accordance with their strictly fortuitous contributions to or withdrawals from survival and reproduction.

However, a single new variant (produced by recombination or mutation), no matter how adaptive, is likely to be swamped in its effects if it appears in a large population. One long-necked giraffe in a million is just not going to make a difference. To begin with, though it can reach food that other animals cannot, it just might be hit by lightning and die before breeding. For another, when the one gene for long necks combines with any of the million or so genes for short necks, the result may just be short necks. For the long-neck genes to make a difference, the number of giraffes with which its bearer breeds must be small, so that copies of the gene in subsequent generations have a chance to combine with one another and produce more long necks. Thus, the structure of the evolving population is important for the occurrence of adaptive evolution: It should be small and make for a certain amount of interbreeding. But if it is too small, a well-adapted population could be wiped out by random forces before it has a chance to expand its numbers.

In addition, for adaptive evolution the environment must remain relatively constant over long periods of time. The environment presents organisms with survival problems. But it takes a generation for the best solutions to these problems to make a difference for the species, for it is only in the relative proportions of the offspring that the best solutions to an adaptational problem have their evolutionary effects. So if an environment changes at rates faster than the generation time of a species, that species will never show any pattern of adaptation. Among animals this is not a serious problem. Most environmental problems – heat, cold, gravity, lack of food in winter, and so on – have been with us for literally geological epochs. And even the generation time of the tortoise – a hundred and fifty years or so – is nothing compared with such epochs. So there has been time and enough stability for a lot of evolution. Nevertheless, critics of Darwinian theory complain that though there has been enough time for the evolution of species, there seems little evidence of transitional forms of the sort we should expect. Indeed, it is an old saw among paleontologists that the fossil record shows mainly that evolution took place elsewhere. Another important thing to note is that in rapidly changing environments, survival puts a premium on generalists who are moderately well adapted to a number of environments over specialists very well adapted to just one sort of environment.

In selecting variants for differential survival, nature works with what variation presents itself and shapes available properties. Thus, it has the appearance of seeking the quick and dirty solution to an adaptive problem, not the optimally adaptive one. Because an organism can make no contribution to evolution unless it survives, nature will work with what is presented to it and will encourage early approximate solutions over late but elegant and exact ones. By the time an elegant solution is available, the lineage may be extinct.

Two other important features of Darwinian evolution are its commitment to gradualism and to individual selection. Neither Darwin nor the majority of contemporary evolutionary biologists believe that evolution has proceeded or can proceed by large changes. Rather, they view evolution as the accretion of large numbers of very small changes over long periods. It is not that great improvements in adaptation over small numbers of generations are impossible. Rather, the evidence such as Darwin understood it – and the genetic mechanism of heredity such as modern geneticists understand it – make such evolutionary jumps highly improbable. Similarly, Darwin shares with modern biologists a conviction that the locus of selection is the individual organism, not larger groups in which individuals participate. If groups of various kinds evolve, then it will be because of the adaptational advantages they accord to individuals who maximize fitness. Groups that disadvantage some of their individual members to increase average fitness, for example, are vulnerable to free riders, who take advantage of benefits groups provide their members while failing to contribute to the provision. These free riders will prosper at the expense of contributors until they have completely displaced them. For groups without enforcement mechanisms, individual selection will always swamp group selection. Moreover, it is hard to see how enforcement mechanisms can emerge in the first place, given individual fitness maximization.

Another feature of evolutionary theory is well worth understanding: its relatively weak powers of prediction. About the only place where there is very strong predictive evidence for natural selection is in laboratory experiments and in what animal and plant breeders call artificial selection. In the lab and on the farm, we can control environmental conditions (reproductive opportunities) stringently enough so that only a narrow class of animals or plants survive and reproduce. The result is relatively rapid changes in the proportions of properties adaptive to our interests as breeders. However, not only have we not produced anything that all will agree constitutes a new species, but, as noted, the fossil record does not help either. Evolutionary biology has no striking retrodiction to its credit, and such predictions as it might make are either very generic or likely to be no more reliable than the initial or boundary conditions to which the theory is applied.

In fact, for much of the century the theory of natural selection has often been stigmatized as completely lacking in evidential bearing, as being an unfalsifiable trivial tautology. The charge is well understood. The theory asserts that the fittest survive and reproduce differentially. But the only applicable uniform quantitative measure of fitness is reproductive rates. Accordingly, the formula becomes those with the highest reproductive rates have the highest reproductive rates. It is therefore no wonder that no evidence can be found that contradicts the theory, nor can we expect to find evidence that strikingly confirms it either. It is no good defending evolutionary theory by rejecting the demand the theory be falsifiable. To do so is just to blame the messenger; for even if strict falsifiability is too stern a test for a scientific theory, it is still a serious weakness of any theory if we cannot identify its causal variables independently of the effects they bring about. And this is indeed the problem for evolutionary theory. A better response to this complaint against the theory is to admit that in general we cannot enumerate what fitness consists in – there are too many determinants of evolutionary fitness to be mentioned in a theory – or whether and how much a property conduces to fitness depends on the environment. And the only thing all determinants of fitness differences have in common is their *effect* on rates of reproduction. So it is natural to measure fitness differences in terms of their common effects. Once we are clear on the difference between fitness and what we use to measure it, the claim that the fittest survive and reproduce in higher numbers is no more vacuous than the claim that increases in heat make thermometers rise. (For a discussion of the predictive power of the theory of natural selection and the charge that it is a vacuous tautology see Rosenberg, 1985.)

There are many ways in which organisms can adapt in response to a given environmental constraint. An ice age presents survival problems that can be solved by growing fur, adding layers of fat, changing shape to minimize surface area, migration or hibernation, and so on. And there are many ways in which an environment can change: temperature, humidity, wind, pressure, flora, fauna, CO_2 concentration, and the like. Multiplying the environmental changes times the number of different adaptational responses to each change makes it clear that interesting generalizations about adaptation are not to be found in the expression of the theory itself. In fact, because of this the theory has pretty much taken the form of stochastic models of changes in gene frequencies. By making certain assumptions about the independence of genes (and therefore observable traits of organisms – phenotypes) from one another and adding assumptions about differences in fitness, size of inter-breeding population, and the like, the biologist can derive conclusions about the change in gene frequencies over time. The question then becomes whether there are biological phenomena that realize the assumptions of the model well enough so that its consequences can guide our expectations about the phenomena. Instead of seeking general laws about the way in which environmental changes result in adaptations, evolutionary biologists consider which models of changes in gene frequencies most clearly illuminate processes of interest and whether the most illuminating models have interesting features in common. By and large, the number of such predictively powerful models has not been great, and they have had relatively few distinctive features in common. This should be no surprise, for if the models were very successful and had a good deal of structure and a large proportion of assumptions in common, then the most obvious explanation of these facts would be the truth of a simple and powerful theory that unified them all and explained why they worked so well. Such a theory would in fact replace the theory of natural selection whose weakness and lack of predictive content leads biologists to seek models of restricted phenomena instead of general laws.

2 ECONOMISTS AND EVOLUTION

Why should anyone suppose that Darwinian evolutionary theory will provide a useful model for how to proceed in economics? One apparently attractive feature of the theory for economists is the methodological defense it seems to provide neoclassical theory in the face of charges that the theory fails to account for the actual behavior of consumers and producers. Thus, Friedman offers the following argument for the hypothesis that economic agents maximize money returns:

> Let the apparent immediate determinant of business behavior be anything at all, habitual reaction, random chance, or whatnot. Whenever this determinant happens to lead to behavior consistent with rational and informed maximization of returns, the business will prosper, and acquire resources with which to expand; whenever it does not, the business will tend to lose resources and can

be kept in existence only by the addition of resources from outside. The process of "natural selection" thus helps to validate the hypothesis. Or, rather, given natural selection, acceptance of the hypothesis can be based on the judgment that it summarizes appropriately the conditions for survival. (Friedman, 1953, p. 35)

This argument does reflect a feature of evolutionary theorizing, though admittedly a controversial one. The natural environment sets adaptational problems that animals must solve to survive. The fact that a particular species is not extinct is good evidence that it has solved some of the problems imposed upon it. This fact about adaptational problems and their solutions plays two roles in evolutionary thinking. First, examining the environment, biologists might try to identify the adaptational problems that organisms face. Second, focusing on the organism, biologists sometimes attempt to identify possible problems that known features of the organism might be solutions to. The problem with this approach is the temptation of Panglossianism: imagining a problem to be solved for every feature of an organism we detect. Thus, Dr. Pangloss held that the bridge of the nose was a solution to the adaptational problem of holding up glasses. The problem with inferences from the environment to adaptational problems is that we need to determine all or most of the problems to be solved, for each of them is an important constraint on what will count as solutions to others. Thus, having a dark color will not be a solution to the problem of hiding from nocturnal predators unless the organism can deal with the heat that such a color will absorb during the day. However, a color that will effect the optimal compromise between these two constraints may fail a third one, say being detectable by conspecifics during mating season.

Then there is the problem of there being more than one way to skin a cat. Even if we can identify an adaptational problem and most of the constraints against which a solution can be found, it is unlikely that we will be able to narrow the range of equally adaptive solutions down to just the one that animals actually evince. Thus, we are left with the explanatory question of why this way of skinning the adaptational cat emerges and not another one apparently equally good. There are two answers to this question. One is to say that if we knew all the constraints we would see that the only possible solution is the actual one. The other is to say that there are more than one equally adequate solutions and that the one finally "chosen" appeared for nonevolutionary causes. The first of these two replies is simply a pious hope that more inquiry will vindicate the theory. The second is in effect to limit evolutionary theory's explanatory power and deny it predictive power.

These problems have in general hobbled "optimality" analysis as an explanatory strategy in evolutionary biology. Many biologists find the temptations of Panglossianism combined with the daunting multiplicity of constraints on solutions to be so great that they despair of providing an evolutionary theory that contributes to our detailed understanding of organisms in their environments.

The same problems bedevil Friedman's conception and limit the force of his conclusion. The idea that rational informed maximization of returns sets a necessary and/or sufficient condition for long-term survival in every possible economic environment, or even in any actual one, is either false or vacuous. Is the hypothesis that returns are maximized over the short run, the long run, the fiscal year, or the quarter? If we make the hypothesis specific enough to test, it is plainly false. Leave it vague and the hypothesis is hard to test. Suppose we equate the maximization of returns hypothesis with the survival of the fittest hypothesis. Then nothing in particular follows about what economic agents do and how large their returns are, any more than it follows what particular organisms do and how many offspring they have. However many the offspring and however much the returns, the results will be maximal, given the circumstances, over the long run. What we want to know is which features of organisms increase their fitness, and which strategies of economic agents increase their returns. And we want this information both to explain particular events in the past and to predict the course of future evolution. For the hypothesis of maximization of returns to play this substantive role, it cannot be supposed to be on a par with the maximization of fitness hypothesis. Rather, we need to treat it as a specific optimal response to a particular environmental problem, rather as we might treat coat color as an optimal response to an environmental problem of finding a color that protects against predators, does not absorb too much heat, is visible to conspecifics, and the like. But when we think of the maximization of returns hypothesis this way, it is clear that maximizing dollar returns is not a condition of survival in general, either in the long or the short run.

As already noted, nature has a preference for quick and dirty solutions to environmental problems. It seems to satisfice, in Simon's phrase. But unlike satisficing, nature's strategy really is a maximizing one. It is just that the constraints are so complicated and so unknown to us that the solutions favored by selection look quick and dirty to us. If we knew the constraints, we would see that they are elegant and just on time. Learning what the constraints are and how the problems are solved is where the action is in vindicating the theory of natural selection, because only this will enable us to tell whether the solution

179

really maximizes fitness, as measured by offspring. Similarly, in economics the action is in learning the constraints and seeing what solutions are chosen. Only this will tell us whether dollar returns are really maximized and whether maximizing dollar returns ensures survival. To stop where Friedman does is to condemn the theory he sets out to vindicate to the vacuity with which Darwinian theory is often charged. If the theory of natural selection is to vindicate economic theory or illuminate economic processes, it will have to do more than just provide a Panglossian assurance that whatever survives in the long run is fittest. What is needed in any attempt to accomplish this is a better understanding of the theory of natural selection. Such an improved understanding of the theory is evident in Alchian's (1950) approach to modeling economic processes as evolutionary ones.

Alchian's approach is not open to obvious Panglossian objections, nor does it make claims about empirical content that transcend the power of an evolutionary theory to deliver. Still, its problems reveal more deeply the difficulties of taking an evolutionary approach to economic behavior.

To begin with, Alchian's approach reflects the recognition that Darwinian theory's claims about individual responses to the environment are hard to establish, impossible to generalize, and therefore without predictive value for other organisms in other environments. Alchian recognized that the really useful versions of evolutionary theory are those that focus on populations large enough that statistical regularities in responses to environmental changes can be discerned. And he recognized that Darwinian evolution operates through solutions to adaptational problems that are, in appearances at any rate, quick and dirty, approximate and heuristic, and not rationally and informationally maximizing. Like the biological environment, the economic one need not elicit anything like the maximization of returns that conventional theory requires:

> In an economic system the realization of profits is the criterion according to which successful and surviving firms are selected. This decision criterion is applied by an impersonal market system . . . and may be completely independent of the decision processes of individual units, of the variety of inconsistent motives and abilities and even of the individual's awareness of the criterion. The pertinent requirement – positive profits through relative efficiency – is weaker than "maximized profits," with which unfortunately, it has been confused. Positive profits accrue to those who are better than their actual competitors, not some hypothetically perfect competitors. As in a race, the award goes to the relatively fastest, even if all competitors loaf. . . . [S]uccess (survival) accompanies relative superiority; . . . it may . . . be the result of fortuitous circumstances. Among all competitors those whose particular conditions happen to be the most

appropriate of those offered to the economic system for testing and adoption will be "selected" as survivors. (Alchian, 1950, pp. 213–214)

Alchian also recognizes that adaptation is not immediate and is discernible to the observer only in the change in statistical distributions over periods of time, and recognizes that what counts as adaptive will change as the economic environment does. Alchian uses a parable to illustrate the way that the economic environment shifts the distribution of actually employed choice strategies toward the more rational:

> Assume that thousands of travelers set out from Chicago, selecting their roads completely at random and without foresight. Only our "economist" knows that on but one road there are any gas stations. He can state categorically that travelers will continue to travel only on that road: those on other roads will soon run out of gas. Even though each one selected his route at random, we might have called those travelers who were so fortunate as to have picked the right road wise, efficient, farsighted, etc. Of course we would consider them the lucky ones. If gasoline supplies were now moved to a new road, some formerly luckless travelers again would be able to move; and a new pattern of travel would be observed, although none of the players changes his particular path. The really possible paths have changed with the changing environment. All that is needed is a set of varied, risk-taking (adoptable [*sic*]) travelers. The correct direction of travel will be established. As circumstances (economic environment) change, the analyst (economist) can select the type of participants (firms) that will now become successful; he may also be able to diagnose the conditions most conducive to greater probability of survival. (Alchian, 1950, p. 214)

To ensure survival and significant shifts in the direction of adaptation, several other conditions must be satisfied: To begin with, the environment must remain constant long enough so that those strategies more well adapted to it than others will have time to out-compete the less well adapted and to increase their frequency significantly enough to be noticed. Moreover, the initial relative frequency of the most well adapted strategy must be high enough so it will not be stamped out by random forces before it has amassed a sufficient advantage to begin displacing competitors. And of course, it must be the case that there are significant differences among competing strategies. Otherwise, their proportions at the outset of competition will remain constant over time. There will be no significant changes in proportions to report.

What kind of knowledge will such an economic theory provide? Even at his most optimistic, Alchian was properly limited in his expectations. He made no claims that with an evolutionary approach the course of behavior of the individual economic agent could be predicted. Here the parallel with evolution is obvious. Darwin's theory not only has no implications for what will happen to any individual organism; its implications for large numbers of organisms are at best probabilistic:

> A chance dominated model does not mean that an economist cannot predict or explain or diagnose. With a knowledge of the economy's requisites for survival and by a comparison of alternative conditions, he can state what types of firms or behavior relative to other possible types will be more viable, even though the firms themselves may not know the conditions or even try to achieve them by readjusting to the changed conditions. It is sufficient if all firms are slightly different so that in the new environmental situation those who have their fixed internal conditions closer to the new, but unknown optimum position now have a greater probability of survival and growth. They will grow relative to other firms and become the prevailing type, since survival conditions may push the observed characteristics of the set of survivors towards the unknowable [to them] optimum by either (1) repeated trials or (2) survival of more of those who happen to be near the optimum – determined ex post. If these new conditions last "very long," the dominant firms will be different ones from those which prevailed or would have prevailed under the other conditions. *Even if* the environmental conditions cannot be forecast, the economist can compare for given alternative potential situations the types of behavior that would have higher probability of viability or adoption. If explanation of past results rather than prediction is the task, the economist can diagnose the particular attributes which were critical in facilitating survival, even though individual participants were not aware of them. (Alchian, 1950, p. 216; emphasis added)

As a set of conditional claims, most of what Alchian says about the explanatory and predictive powers of an evolutionary theory of economic processes is true enough. The trouble is that almost none of the conditions obtain, either in evolutionary biology or in economic behavior, that would make either theory as useful as Alchian or any economist needs it to be. Thus, the attractions of an evolutionary theory for economists must be very limited indeed. Alchian rightly treats the economy as the environment to which individual economic agents are differentially adapted. As with the biological case, we need to know what "the requisites of survival" in the environment are. In the biological case, this is not a trivial matter; and beyond the most obvious adaptational problems, there are precious few generalizations about what any particular ecological environment requires for survival, still less what it rewards in increased reproductive opportunities. We know animals need to eat, breathe, and avoid illnesses and environmental hazards, and the more of their needs they fulfill the better off they are. But we don't know what in any given environment the optimal available diet is or what the environmental hazards are for each of the creatures that are its inhabitants. And outside of ecology and ethology, few biologists are interested in this information in any case, for its systematic value to biology is very limited. Ignorance about these requisites for survival in biology make it difficult to predict even "the types of . . . behavior relative to other possible types that will be more viable." It is easy to predict that all surviving types will have to subsist in an oxygen-rich environment where the gravitational constant is 32 feet/second and the ambient temperature ranges from 45 degrees to minus 20 degrees Celsius. But such a "prediction" leaves us little closer to what we hope to learn from a prediction. The same must be true in evolutionary economics. We have no idea of what the requisites for survival are, and even if we learned them, they would probably not narrowly enough restrict the types that can survive for us to frame any very useful expectations of the future. Of course, this is not an in-principle objection to an evolutionary approach. But consider what sort of information would be required to establish a very full list of concrete, necessary conditions for survival of, say, a firm in any very specific market, and then consider the myriad ways in which economic agents could so act to satisfy those conditions. This information is either impossible to obtain or else, if we had it, an evolutionary approach to economic processes would be superfluous. To see this, go back to Alchian's discussion.

Alchian notes that over time the proportion of firms of various types should change: The proportion of those that are more fit should increase while those less fit should decrease. If environmental conditions last a long time, "the dominant firms will be different from those which prevailed . . . under other conditions." True enough, but what counts as a long time for environmental conditions? In the evolutionary context, "long enough" means at least one generation, and the duration of a generation will vary with the species. In addition the notion of long enough reflects a circularity that haunts evolutionary biology. Evolution occurs if the environment remains constant long enough for the proportion of types to change. Long enough is enough time for the proportions to change. Moreover, when the number of competing individuals is small, there may be change in proportions of types that is not adaptational, but is identified as drift – a sort of sampling error. But what is a small number of individuals versus a large number? Here, the same ambiguity emerges. Large enough means a number in which changes in proportion reflect evolutionary adaptation. The only way in which to break out of this circularity of long enough, large enough, and so on is to focus on individual populations in particular environments over several generations. And the answer we get for any one set of individuals will be of little value when we turn to another set

183

of the same types in different environments or to different types in the same environment.

Can the situation be any better in economics? In fact, won't the situation be far less promising? After all, the environment within which an economic agent must operate does not change with the stately pace of a geological epoch. Economic environments seem to change from day to day. If they do, then there is never enough time for the type most adapted to one environment to increase in its proportions relative to other types. Before the type has had a chance to do so, the environment has changed, and another type becomes most adapted. But perhaps economic environments do not change quite so quickly. Perhaps to suppose that they do change so quickly is to mistake the weather for the climate. Day-to-day fluctuations may be a feature of a more long-standing environment. The most well adapted individual to an environment is not one who responds best to each feature of it, including its variable features, but rather one who adapts best overall on an average weighted by the frequency with which certain conditions in the environment obtain. So the period of time relevant to evolutionary adaptation might be long enough for changes in proportion to show up. For the parallel to evolution to hold up, this period of environmental constancy will have to be longer than some equivalent to the generation time in biological evolution. But is there among economic agents any such an equivalent? Is there a natural division among economic agents into generations? With firms, the generation time might be the period from incorporation to the emergence of other firms employing the same method in the same markets through conscious imitation; with individual agents, the minimal period for evolutionary adaptation will be the time during which it takes an individual to train another to behave in the same way under similar economic circumstances. But these two periods are clearly ones during which the economic environment almost always changes enough to shift the adaptational strategy.

The only way we can use an evolutionary theory to predict the direction of adaptation is by being able to identify the relevant environment that remains constant enough to force adaptational change in proportions of firms. As Alchian tacitly admits, this is something we cannot do: "Even if the environmental conditions cannot be forecast, the economist can compare for given alternative potential situations the types of behavior that would have a higher probability of viability or adoption" (p. 217). This is a retrospective second best. Suppose economically relevant environmental conditions could be forecast. Then, it is pretty clear we would not need an evolutionary theory of economic behavior. Friedman's rationale for neoclassical theory would then come into its own. If we knew environmental conditions, then we could state what optimal adaptation to them would be. And if we could do this, so could at least some of the economic agents themselves. To the extent that they could pass on this information to their successors, Panglossianism would eventually be vindicated in economic evolutionary theory. Economic agents would conform their actions to the strategy calculated to be maximally adaptive, just as Friedman claims. An evolutionary theory of economic behavior is offered either as an alternative to rational maximizing or as an explanation of its adequacy. If rational maximizing is adequate as a theory, evolutionary rationales are superfluous; if it is not adequate, then an evolutionary approach is unlikely to be much better, and for much the same reason: Neither economic theorists nor economic agents can know enough about the economic environment for the former's predictions or the latter's decisions to be regularly vindicated.

3   WHY ECONOMICS IS NOT DARWINIAN

One of the features of evolutionary theory that makes it attractive to the economist is the role of equilibrium in claims made about nature. Equilibrium is important for economic theory not least because of the predictive power it accords the economist. An economic system in equilibrium or moving toward one is a system some or all of whose future states are predictable by the economist. Equilibrium has other (welfare-theory-relevant) aspects, but its attractions for economists must in part consist in the role it plays so successfully in physical theory and evolutionary theory. Evolutionary biology defines an equilibrium such that gene ratios do not change from generation to generation, and it stipulates several conditions that must obtain for equilibrium: a large population mating at random, without immigration, emigration, or mutation, and of course, without environmental change. Departures from these conditions will cause changes in gene frequencies within a population. But over the long run, the changes will move in the direction of closer adaptation to the environment, either closer adaptation to an unchanged one or adaptation toward a new one. The parallel to economic equilibrium is so obvious that mathematical biologists have simply taken over the economist's conditions for the existence and stability of equilibria. If a unique, stable market-clearing equilibrium exists, then its individual members are optimally adapted to their environment, no trading will occur, and there will be no change – no evolution – in the economy. But if one or another of the conditions for equilibrium is violated, an efficient economic system will either move back to the original equilibrium or to a new one by means of adjustments in which individuals move along paths of increased adaptation.

In evolutionary biology, equilibrium has an important explanatory role. As far as we can see, populations remain fairly constant over time, and among populations the proportions of varying phenotypes remain constant as well. Moreover, when one or another of the conditions presupposed by equilibrium of gene frequencies is violated, the result is either compensating movement back toward the original distribution of gene frequencies or movement toward a new level of gene frequencies. These facts about the stability of gene frequencies and their trajectories need to be explained, and the equilibrium assumptions of transmission genetics are the best explanations going. In addition they will help us make generic predictions that when one or another condition, like the absence of mutation, is violated, a new equilibrium will be sought. Sometimes we can even predict the direction of that new equilibrium. But in real ecological contexts (as opposed to simple textbook models), we can hardly ever predict that actual value of the new equilibrium level of gene frequencies. This is because we do not know all the environmental factors that work with a change in one of these conditions, and among those factors that are known, we have only primitive means of measurement for their dimensions.

Now compare economics. To begin with, we have nothing comparable to the observed stability of gene frequencies that needs to be explained. So the principal explanatory motivation for equilibrium explanations is absent. We cannot even appeal to the stability of prices as a fact explaining equilibrium in economics because we know only too well that neoclassical general equilibrium theory has no explanation for price stability. That is, given an equilibrium distribution and a change in price, there is no proof that the economy will move to a new general equilibrium. (For this reason, general equilibrium theory has recourse to the Walrasian auctioneer and *tâtonnement.*)

There is no doubt that economic equilibrium theory has many attractive theoretical features – mathematical tractability, the two welfare theorems – but it lacks the most important feature that justifies the same kind of thinking in evolutionary biology: independent evidence that there is a stable equilibrium to be explained.

One of the factors giving us some confidence that equilibrium obtains with some frequency in nature is that changes in gene frequencies are not self-reinforcing. If some change also affects gene frequencies, then such change in gene frequencies will rarely precipitate still another round of changes in gene frequencies, and so on, thus cascading into a period of instability. Of course, sometimes evolutionary change is "frequency dependent": If one species of butterfly increases in population size because it looks like another species that birds avoid, then once it has grown larger in number than the bad-tasting butterflies, its similar appearance and the genes that code for appearance will no longer be adaptive and may decline. But presumably the proportions will return to some optimal level and be held there by the twin forces of adaptation and maladaptation.

In the game theorist's lingo, evolutionary adaptational problems are parametric: The adaptiveness of an organism's behavior does not depend on what other organisms do. But we cannot expect this absence of feedback in economic evolution. Among economic agents, the problem is strategic. Economic agents are far more salient features of one another's environment than animals are features of one another's biological environment. Changes in agents' behaviors affect their environments regularly because they call forth changes in the behavior of other agents, and these further changes cause a second round of changes in the original agents' behavior. Game theorists have come to identify this phenomenon under the rubric of the common knowledge problem. Economists traditionally circumvented this problem by two assumptions that have parallels in evolutionary biology as well. It is important to see that the parallels do not provide much ground for the rationalization of economic theorizing in the biologist's practice.

Both evolutionary equilibrium and economic general equilibrium require an infinite number of individuals. In the case of evolution, this is to prevent drift or sampling error from moving gene frequencies independent of environmental changes. In the case of the theory of pure competition, it is to prevent agent choice from becoming strategic. If the firm is always a price taker and can have no effect on the market, then it can treat its choices as parametric. Where the number of interactors is small, the assumption of price taking produces very wrong predictions, and there is indeed no stability and typically no equilibrium.

Is sauce for the biological goose sauce for the economic gander? Can both make the same false assumption with equal impunity? The fact is, though assumptions of infinite population size are false for interbreeding populations, it seems to do little harm in biology. That is, despite the strict falsity of the evolutionary assumption, populations seem to be large enough for theory that makes these false assumptions to explain the evident facts of constancy and/or stability of gene frequencies. In the case of economics, there are no such evident facts, and one apparent reason seems to be the falsity of the assumption of an infinite number of economic agents.

The other assumption evolutionary theory and economic theory traditionally make is that the genes and the agents, respectively, are "omniscient." Genes carry information in two senses. First, they carry instructions for the building and maintenance of proteins and assemblies of genes that meet the environment as phenotypes. Second, they indirectly carry information about which phenotypes are most adapted to the environment in which they find themselves. They do so through the intervention of selective forces that cull maladaptive phenotypes and thus the genes that code for phenotypic building blocks. And as long as the environment remains constant, the gene frequencies will eventually track every environmentally significant, biologically possible adaptation and maladaptation. In this sense the genome is in the long run omniscient about the environment. There are two crucial qualifications here. First there is the assumption of the constancy of the environment, something economic theory has little reason to help itself to. Second, there is the "long run" – another concept evolutionary theory shares with economic theory. Evolutionary biology has world enough and time for theories that explain and predict only in the long run – geological epochs are close enough to infinite not to matter for many purposes. But Keynes pointed out the problem for economics concerning theories that explain only the long run. An evolutionary economic theory committed to equilibrium is condemned at best to explain only the long run.

We know only too well the disequilibrating effects of nonomniscience, that is, how information obstructs the economy's arrival at or maintenance of an equilibrium. Indeed, the effects of differences in information on economic outcomes are so pervasive that we should not expect economic phenomena ever to reflect the kind of equilibrium evolutionary biological phenomena do. Arrow has succinctly summarized the impact of information on equilibrium models:

> If nothing else there are at least two salient characteristics of information which prevent it from being fully identified as one of the commodities represented in our abstract models of general equilibrium: (1) it is, by definition, indivisible in its use; and (2) it is very difficult to appropriate. With regard to the first point, information about a method of production, for example, is the same regardless of the scale of the output. Since the cost of information depends only on the item, not its use, it pays a large scale producer to acquire better information than a small scale producer. Thus, information creates economies of scale throughout the economy, and therefore, according to well-known principles, causes a departure from the competitive economy. Information is inappropriable because an individual who has some can never lose it by transmitting it. It is frequently noted in connection with the economics of research and development that information acquired by research at great cost may be transmitted much more cheaply. If the information is, therefore, transmitted to one buyer, he can in turn sell it very cheaply, so that the market price is well below the cost of production. But if the transmission costs are high, then it is also true that there is inappropriability, since the seller cannot realize the social value of the

information. Both cases occur in practice with different kinds of information. But then, according to well-known principles of welfare economics, the inappropriability of a commodity means that its production will be far from optimal. It may be below optimal: it may also induce costly protective measures outside the usual property system. Thus, it has been a classic position that a competitive world will underinvest in research and development, because the information acquired will become general knowledge and cannot be appropriated by the firm financing the research. If secrecy is possible, there may be overinvestment in information gathering, each firm may secretly get the same information, either on nature or on each other, although it would of course consume less of society's resources if they were collected once and disseminated to all. (Arrow, 1984, pp. 142–143)

If agents were omniscient, these problems would not emerge. Successful genomes are omniscient (at least in the long run) so the parallel problems do not emerge in nature, and do not obstruct equilibria. There are no apparent economies of scale operating within species in reproductive fitness. And besides, the information that the environment provides about relative adaptedness is costless and universally available. So there is no problem about appropriability. In the absence of secrecy and the need for strategic knowledge about what other agents know, there is no room in biological evolution for the sort of problems information raises in economics. Once biological systems become social and their interactions become strategic, the role for information becomes crucial. But at this point evolution turns Lamarckian. It is no surprise that when "acquired" characteristics are available for differential transmission, markets for the characteristics will emerge. But at this point Darwinian evolution is no longer operating. In fact, one good argument against the adoption of Darwinian evolutionary theory as a model for economic theory is just the difference made by information. Once it appears in nature, evolution ceases to be exclusively or even mainly Darwinian. Why suppose that once information becomes as important as it is in economic exchange that phenomena should again become Darwinian?

4   THE CASE OF NELSON AND WINTER

It will be useful to apply some of the animadversions already advanced to what is doubtless the most well developed approach to economics inspired by Darwinian considerations (Nelson and Winter, 1982). The capstone of two distinguished careers, few books can have had a more disappointing reception in recent economics. I hazard two guesses about why it has fallen stillborn from the presses, both of them natural consequences of what I have argued here. First, the predictive power of Nelson and Winter's evolutionary alternative to neoclassical theory is by their own admission little better than that of the theory of natural selection. And for Nelson and Winter, as for most other economics, predictive improvement is an important criterion of theoretical advance. But predictive power is the least of variation and selection theory's virtues, no matter what discipline it is developed for. No one should have expected more in application elsewhere. Second, Nelson and Winter's theory doesn't really look very different from neoclassical theory: Like other evolutionary theories, it lends itself to equilibria explanations, it reflects the fact that biological evolution fosters constrained maximization, and its interests are in the aggregate of individuals and not in their particular behavior. Except in areas where neoclassical theory is silent, the evolutionary approach is not different enough to make it more than a variation on the theme written in the eighteenth century by Adam Smith.

But once we get past the obvious features of Nelson and Winter's theory, there is a more fundamental problem that deprives it, I fear, of even the prospects of being as good as Darwinian evolutionary theory. If an evolutionary approach to economics is to be something more than a suggestive metaphor, if we are actually to confirm the claim that economic phenomena reflect a process of Darwinian evolution, we must identify in the phenomena the fundamental causal forces that are necessary and sufficient for natural selection. Because it fails to do this Nelson and Winter's theory is at best a metaphorically Darwinian theory that stands or falls on its own, without the support or the drag of Darwin's theory.

Evolutionary biologists and philosophers of biology have identified the following minimal condition for evolution by natural selection: There must be *replicators* and there must be *interactors* (see Dawkins, 1976; Hull, 1989). A replicator is defined as an entity that passes on its structure intact in successive replications. Thus, a gene is a replicator, but not the only possible replicator. If organisms produce offspring very much like themselves in structure, then organisms are replicators too. The key to deciding whether something is a replicator is what counts as a distinct offspring whose structure can be compared. An interactor is an entity that interacts as a cohesive whole with its environment in such a way that this interaction causes replication to be differential. Organisms are paradigm interactors in evolutionary biology. But they are not the only possible ones: Genes, cells, tissues, and organs interact with their bodily environments in ways that cause more or less copies of the genes to be produced. In fact, some evolutionary biologists have in the past suggested that groups, populations, and even whole species might constitute interactors. This view is no longer widely held (for reasons well understood by game theorists; see Smith 1984). Selection is the process in which differential extinction and proliferation of interactors cause the differential perpetuation of the relevant replicators. Some evolutionary biologists define one other term, the *lineage*, which is the entity that actually evolves, as the proportion of types of interactors in its line of descent changes from generation to generation. It is important to bear in mind that interactors do not evolve; only the lines of descent of which they are composed do.

Nelson and Winter's aim is not simply to wrap economic theory in the mantle of evolution, but to show how economic processes instantiate Darwinian natural selection. Though they do not use the word, they identify the replicator as a behavioral routine.

> At any time a firm's routines define a list of functions that determine . . . what a firm does as a function of various external variables (principally market conditions) and internal state variable (for example the firms prevailing stock of machinery or the average profit rate it earned in recent periods) (p. 16). Organizational capabilities consist largely of the ability to perform and sustain a set of routines; such routines could be regarded as a highly structured set of "habitual reactions" linking organization members to one another and to the environment. The tendency for such routines to be maintained over time plays in our theory the role that genetic inheritance plays in the theory of biological evolution. (p. 142)

The interactors are pretty clearly meant to be firms. Here is a typical expression of this commitment, as Nelson and Winter construct the parallels:

> The comparative fitness of genotypes (profitability of routines) determines which genotypes (routines) will tend to become predominant over time. However the fitness (profitability) clearly depends on the characteristics of the environment (market prices) confronting the species (collections of firms with similar routines). The environment (price vector) in turn depends, however, on the genotypes (routines) of all the individual organisms (firms) existing at a time – a dependency discussed in the subdiscipline called ecology (market theory). (p. 160)

So replicator = behavioral routine, of which there are several different kinds; interactor = firm; fitness is measured in profits; and the environment is given by the price vector. But then what evolves? What is the lineage, what are the generations, what is the principle by which we individuate members of the lineage to establish intergenerational selection? Here, incoherence sets in, because Nelson and Winter also identify the unit of evolution as the firm – in other words, in their model the interactor is both the lineage of firms and its proper parts, the individual firms, for it is individual firms that grow in size and evolve, like populations. Nelson and Winter write, "Through the joint action of search [for routines] and selection, the firms evolve over time" (p. 19). Thus, firms are both interactors and lineages. But if they are lineages, they must be composed of interactors as well.

At a minimum, the neatness of evolutionary arithmetic is destroyed by this complication. More likely, the result is the surrender of a Darwinian approach to economic evolution in favor of a Lamarckian one, in which anything can evolve into anything by any means. In the biological case to determine evolution within the lineage, one need only count the proportions of the distinct types of its component individual members. Changes that take place within an individual member of a lineage are matters of development. From a Darwinian perspective, these changes cannot count as evolution or even have a role in it. Only differential reproduction of the replicators – the genes or routines that give rise to them – can do this. The only way changes in interactors can themselves have an effect on evolution of the lineage is by Lamarckian means – the inheritance of acquired characteristics. Lamarckianism is, of course, not the label for an alternative evolutionary theory. It is just a label for the claim that change is not Darwinian. As such it sheds no special light on economic processes, or any other.

Might we preserve the Darwinian character of Nelson and Winter's theory by a little reorganization? Let's try. By the adoption of more adapted routines, a firm develops, grows in size, attains profitability, and the like. But if the firm also evolves, then it must at least be (one of) its own descendant(s), otherwise the analogy with evolutionary theory breaks down. For this sort of change to also count as evolution, the improved adaptation must be transmitted to the firm's descendants, successor, subsidiaries, and so on. But what if the firm becomes so adapted that it swamps the competition, spins off no subsidiaries, and grows to become a natural monopoly? Are we to say that the lineage of which it is a member has evolved in the direction of greater adaptation? The number of members of the lineage continually decreases until there is just one member left, and it is now vulnerable to extinction in a sudden environmental change. A reasonable thing to say about this scenario is that the single firm left has shown itself to be the fittest. Reasonable, but not from the perspective of Darwinian evolution and not just because only one individual is left at the end of a period of evolution. This way of describing the outcome of the evolutionary process obscures a real Darwinian insight.

An evolutionarily coherent version of Nelson and Winter makes routines be replicators and the organizational unit that employs the routines the interactor. If the routine is a marketing strategy, the interactor is the smallest marketing department that can execute it; if the routine is a production technique, the interactor is the smallest shop-floor team that employs it. Each of these will grow in size subject to the constraints of other interactors, whether in the same or other firms. Firms are sets of interactors that are coadapted, that work together to produce outcomes that increase their numbers, either within a firm, if it can keep the routines proprietary, or across an industry. On this view firms are not lineages either. Lineages are composed of the smallest organizational units that reproduce. Firms cut across lineages, bundling together varying numbers of varying organizational units. Changes in the size and profitability of firms will reflect adaptedness of their component units to one another and to the industrial environment. Adaptation of interactors to the environment may result in rapid increase of the number of firms in an industry, each of them composed of a small number of units, or a decrease in the number of firms, each bundling together a vast number of the same units. Compare the industry structure for PC clones with the industry structure for supercomputers. From the point of view of natural selection, tracking the fiscal growth or shrinkage of firms may reveal little about the way in which Nelson and Winter's replicators determine economic evolution. If this is in fact the way to apply evolutionary theory to the firm's behavior, the result will be a discipline more like organization theory than economics. Few economists (besides March, Cyert, and Simon) want this. The evolutionary biologist will not be surprised, for organization theory looks like something biologists do. One important component of the biologist's discipline is empirical work in the field – taxonomy to identify the interactors and ecology to establish the communities of coadapted interactors and to identify the environmental forces that shape them and their adaptations. The purely theoretical modeling comes much later and has little additional explanatory or predictive power. But this is a division of labor that few economists will volunteer for. After all, theirs is a discipline that prescinds from psychology, sociology, and other details.

Of course, if we don't like the turn that our evolutionary theory has taken, we can go back to the drawing board and select new replicators, which will give us firms as the desired interactors and reflect a parallel with evolutionary theory. But now the tail is wagging the dog. We are rearranging our theory so that it will look like one that works well elsewhere. But what is the point?

The role of metaphors in science is not well understood. Indeed, the role of metaphors is still controversial on its home ground in language. It should be no surprise that when we metaphorically or otherwise extend literary metaphor to scientific practice, matters become quickly obscure. Darwin's notion of blind variation and natural selection has been one of the most tempting of metaphors in the social sciences. Whether it has been a source of fruitful stimulation is debatable. Whether the theories cut to its pattern have been well confirmed or not seems to me to be the only interesting question for scientists in these disciplines to actually concern themselves with. All the rest is ad hominem argument or the genetic fallacy.

NOTE

Reprinted from P. Miroski (ed.), *Natural Images in Economic Thought* (Cambridge: Cambridge University Press, 1994), pp. 384–407.

10

The Political Philosophy of Biological Endowments

Some Considerations

Is a government required or permitted to redistribute the gains and losses that differences in biological endowments generate? In particular, does the fact that individuals possess different biological endowments lead to unfair advantages within a market economy? These are questions on which some people are apt to have strong intuitions and ready arguments. Egalitarians may say yes and argue that as unearned, undeserved advantages and disadvantages, biological endowments are never fair, and that the market simply exacerbates these inequities.[1] Libertarians may say no, holding that the possession of such endowments deprives no one of an entitlement and that any system but a market would deprive agents of the rights to their endowments.[2] Biological endowments may well lead to advantages or disadvantages on their view, but not to unfair ones.

I do not have strong intuitions about answers to these questions, in part because I believe that they are questions of great difficulty. To begin, alternative answers rest on substantial assumptions in moral philosophy that seem insufficiently grounded. Moreover, the questions involve several problematical assumptions about the nature of biological endowments. Finally, I find the questions to be academic, in the pejorative sense of this term. For aside from a number of highly debilitating endowments, the overall moral significance of differences between people seems so small, so interdependent and so hard to measure, that these differences really will not enter into practical redistributive calculations, even if it is theoretically permissible that they do so.

Before turning to a detailed discussion of biological endowments and their moral significance, I sketch my doubts about the fundamental moral theories that dictate either the impermissibility or the obligation to compensate for different biological endowments.

195

The fullest egalitarian argument for counterbalancing differences due to natural endowments is provided by Dworkin,[3] and I shall discuss it at length below. But it is worth noting that Dworkin's results rest on the assumption that equality of some kind is morally obligatory, yet Dworkin does not even argue that equality is morally desirable, still less required. And this is typical of many treatments of the demands of equality. As Nozick has noted, however, "it cannot merely be assumed that equality must be built into any theory of justice."[4] I share with Nozick the desire to see a good argument for this presumption.

The best libertarian arguments against equalizing for different endowments are disarmingly simple: people have natural rights, among which are the ownership of their bodies and whatever they acquire and transfer in permissible ways. Advantageous biological endowments fall under the first of these headings and so cannot be expropriated, even to compensate the biologically disadvantaged. To this point the libertarian may, like Nozick, add arguments against versions of the claim that biological endowments are morally arbitrary and therefore should not determine people's "holdings."[5] The trouble with these arguments is that they start too late in the game: they leave unfounded the notion that people have natural rights to anything, including their bodies. Nozick is guilty of a charge of the same gravity as he lodges against egalitarianism: one cannot merely assume that natural rights must be built into any theory of justice.

Yet these notions are crucial to the question about biological endowments in the market. On one side, it is held that only the market can respect the natural rights to endowments that each individual possesses, and that trade which respects these rights is the only fair way, or is the optimal way to distribute the goods and services each agent disposes. On the other hand, it is the inequalities and inefficiencies of markets that lead some to advocate redistribution in the direction of greater equality. Just as monopoly and oligopoly constitute inequalities that destroy the allocative efficiency and welfare-optimizing properties of a market, so too individual differences in biological endowments distort markets unfairly. Accordingly, they must be equalized. Both of these prescriptions accord the market at least the standing of a morally permissible institution, and so will I in what follows.[6] But neither has much force in the absence of prior arguments, either for the existence of natural rights or for the claims of equality in endowments and their effects.

But, as I said, it may turn out that these fundamental moral problems about biological endowments are too academic to be very pressing. To see this, we need to examine the notion of biological endowment itself.

196

I

Three preliminary questions suggest themselves immediately: What exactly are biological endowments? Are there significant differences among them? Do such differences have moral ramifications? The first two of these questions have been pretty widely ignored by those who have discussed this subject, in the strong conviction that the nature of such endowments is clear and their existence beyond any doubt. Who could deny that there are talents and disabilities? And these are just what we call advantageous biological endowments and disadvantageous ones. (A semantic digression: I say "disabilities" instead of "handicaps," which suggests that disabled persons have their caps in hand.) I believe that the first two questions are complex, and that the complexities always affect, and sometimes will bedevil, answers to the third question.

Let us begin with the problem of what a biological endowment is. There may be a presumption that the kind of biological endowments at issue here are hereditary ones, genetically programmed traits that are immutable and inevitable. The trouble with this conception is that there is no biological endowment of *Homo sapiens* that is *purely* under the control of the genome, not even the DNA's own molecular structure. All so-called hereditary endowments are the result of interaction between the genome and its environment. A simple example is PKU disease, an inability to metabolize phenylalanine, which is caused by a small number of defects in the genome and results in mental retardation. But this effect obtains only when phenylalanine is present in the diet. Where it is absent, through planning or accident, there is no such disability. In general the genome interacts with the cellular, tissue, organ, whole body, and ecological environments to produce the kinds of endowments we are interested in – talents and disabilities. In different environments, even slightly different ones, the same gene sequence will produce different traits. And, of course, this result is not restricted only to prenatal environments. The genome interacts with the environment throughout life. To the extent that the environment changes, biological endowments will vary. Moreover, our environment is not just biological: it is social, and it can be manipulated not only by other people, but by the agent himself. Thus, some biological endowments will not be under any agent's control, others will be the responsibility of those who nurture the agent, and still others will be the responsibility of the agent himself.

Let us distinguish these three types of endowments as, respectively, hereditary traits, domestic traits, and acquired traits. Examples of each of these may be helpful. Exceptional eyesight is a hereditary trait on this scheme. It may not be coded by the genome, but it is produced by the genome's interaction with

197

the environment in ways we do not understand and do not attempt to control (as yet). A talent for the violin is, on the other hand, presumably the result of genetic factors and a great deal of parental intervention ("Practice, practice!") at an age before the child can take credit for developing the talent in question. Finally, the adolescent's scholarship-winning talent for basketball is often developed as a combination of genetic endowment and long hours of practice chosen freely by the young person in question, without the encouragement, sacrifice, or even approval of any adult.

Similarly, we may identify disabilities of all these three types. Sickle-cell anemia is a good example of a hereditary disability. PKU disease can be the result of parental failure to monitor diet. Permanent brain damage may be the result of choosing to ride motorcycles without a helmet.

Now we may ask the following question: do all or any of these three types of traits generate unfair advantages or disadvantages in a market economy, or in any other type of social arrangement? That they provide advantages and disadvantages is evident. That they should all be the bases for special charges or compensations is less so. One plausible view is that "society" should compensate individuals for hereditary disabilities, their parents should do so for domestic disadvantages, and no one is obliged to succor those who voluntarily disable themselves. The reasons reflect our beliefs about where responsibility is to be attributed for the causes of these disadvantages. Similarly, it may be thought, self-made talents should not be "taxed" the way we might deal with large inheritances. Those advantaged by the sacrifice of parents or others owe them a debt, but they do not owe "society" one. And those endowed with seven feet of height plus coordination and reflexes seem more subject to redistribution of their gains than others. The reason behind these intuitions appears to be that acquired talents and disabilities are "earned" or "deserved" while hereditary traits are not, and domestic traits are somewhere between the two extremes.[7]

We will have to take this view seriously if it turns out that most of the talents and disabilities that affect (dis)advantages in the market are of the acquired or domestic sort. If this is the case, then the revenue from any charge on undeserved hereditary endowments would be small, and the total transfer-expenditure on compensation for disadvantageous ones might also be small. What would make the issue of unfairness serious is the notion that, earned or not, biological endowments are subject to charges and compensations. Without it, the egalitarian claim loses much of its bite. Moreover, to admit that many or most inequalities are earned or deserved, and yet still should be expropriated in the interests of equality, requires a very strong argument in favor of equality as an overriding ethical good.

I hazard a guess that many or most of the endowments that make a difference in the market are of the domestic or acquired sort: intelligence, self-discipline, the ability to get on with others. What is clear is that medical technology can in the long term reduce the level of hereditary differences far enough to increase further the proportion of earned to unearned advantages and disadvantages. This will make the need for arguments for equality despite unequal desert still more pressing.

There is another serious complication facing any attempt to identify biological endowments. We have introduced the notions of ta'ent and disability to describe biological endowments. From a biological point of view, a talent will be some feature that is adaptive for the organism, either in the evolutionary sense or in some kindred sense of the term. A disability will be some maladaptive trait. But these traits will have to be divergences from the normal or typical endowments of members of the species, for they are supposed to be differences that produce advantages and disadvantages by comparison to normal individuals. This will require some baseline repertoire of abilities common to normal agents. One trouble with this strategy is that modern biology suggests that variation of many endowments within a species is not a matter of normality and disturbances from it, but of the random distribution of relatively discrete traits. Accordingly, there is no "basic" specimen range of endowments to serve as a benchmark in any such assignment of disabilities and talents. Population averages will not enable us to identify base-line levels of talents and disabilities, for there may be no member of the population that is anywhere near the average even when it makes sense to compute one. In such cases, the average is a statistical artifact that may have no implementable significance. The point is not just that for some average level of a trait no individual may actually have exactly that much. Rather, contemporary biological theory does not identify the average value as the *normal* or *natural* level.[8] There is no such thing. There is, instead, a range of variation in the incidence of a trait on which selection works. The upshot is another serious problem for any attempt to identify biological endowments: we cannot look to a purely positive and relatively well established theory like biology or any of its compartments to identify the base-line level of biological endowments, nor can we look to it for an identification of what counts as an exceptional disability or talent. In fact, a little further biological reflection shows that the problem is even more serious.

Biological endowments are "relational" in at least two respects. As we have seen, a "hereditary trait" is in fact always the result of interaction between the genome and the environment. The same genetic information will produce an utterly different phenotype in only slightly different environments.

Additionally, identifying a biological endowment as advantageous or disadvantageous is also a matter of interconnection between the trait and the environment in which it figures. In and of itself this relational feature of biological endowments is no problem. Biologists recognize that one environment's adaptation is another one's maladaptation. For instance, the hereditary sickle-cell trait is an adaptation in a malarial environment and a maladaptation in a malaria-free one.

But the relational character of disabilities and talents makes a special complication for our question. Whether an agent's endowment is a talent or a disability will hinge on the distribution of other talents, disabilities, and preferences among other agents in the society. Thus, being color-blind will be an advantageous biological endowment where there is a high demand for photo reconnaissance experts, and a seriously disadvantageous endowment in a society where everything is color-coded. Accordingly, no theory will enable us to determine whether certain endowments are advantageous or not without a great detail of information about the society as a whole. This is a serious problem for any centralized determination of charges and compensations for disabilities and talents. To know whether a certain trait is a disability may require an impossibly complicated census of the whole society and a good prediction about the future course of its members' tastes, preferences, and technological possibilities.

What is more, since preferences vary over individuals even in the same circumstances, whether a biological endowment turns out to be a disability or a talent will vary from individual to individual and will hinge on preferences. Color blindness may have been more valuable to those with a preference for draft dodging than to those who yearned to be photo intelligence experts, and still more valuable to them than to the aspiring fighter pilot. Accordingly, the census of talents and disabilities we need must somehow reflect or record the preferences of all members of society. The more dissimilar people's preferences are, the harder it will be to assemble useful aggregate data and employ them to make sensible identifications of what count as talents and disabilities. It will turn out to be difficult to treat people of biologically equal endowments equally. For given differences in their own and other's preferences, one man's talent will be another's disability. We shall return to the problem of the interaction of endowments and preferences below. For the moment it is enough to see that the identification of these endowments as talents or disabilities is a very complex matter. Of course, there are some clear cases of talents and disabilities: blindness, paraplegia, sprinter's speed, artist's creativity, and so forth. But behind the identification of each of them lie theoretical presuppositions of a very complicated nature, complicated enough to reduce our confidence about our ability to identify the incidence and magnitude of talents and disabilities beyond these most evident sorts.

II

Let us turn now to the question of the morally relevant advantages talents and disabilities provide. Usually, the advantages that concern people who deal with this question are advantages of wealth and the disposition of other resources, such as power. But some of the advantages biological endowments provide are not fungible, and some are entirely "monadic." That is, they can advantage only their "bearer," and never at the expense of others. An extreme case might be the production of natural opiates that provide a high level of ambient pleasure, no matter what the external conditions of the agent. These advantages will obtain under any social arrangements. Those who hold that such natural advantages are not unfair, while others are, will have to provide a sound basis for this moral distinction. An unusually high level of endorphins seems no different in its hereditary basis than, say, great height plus coordination. It is equally undeserved. If an agent cannot be "charged" for this undeserved advantage, what exempts it from a principle that unearned advantages are unfair? Is it the mere fact that he does not use the advantage to gain money, but rather some other, more desired benefits?

Though hereafter we shall focus on economic advantages and disadvantages that biological endowments generate, it is worth reminding ourselves that these are not the only sort there are. Dworkin makes this point eloquently:

> The assumption that a life dedicated to the accumulation of wealth or to the consumption of luxuries . . . is a valuable life for people given only one chance to live . . . comes as close as any theory of the good life can to naked absurdity.[9]

The ability to enjoy tranquil contemplation of a pastoral scene, whether inborn or acquired, is an advantage. The rewards of uncompensated scholarship, as we all know, are a nonmonetary advantage provided by our talents.

But, adopting for the moment the narrow view of pecuniary advantage, what sort of advantages and disadvantages do talents and disabilities provide? Let us focus on talents for the moment. To determine the pecuniary advantage of a talent, it is useful to think of talents as *capital goods.* There is something of a theory already available about capital goods. And the question of whether such goods provide unfair advantages in a market economy is certainly a prominent one. Accordingly, we may be able to take advantage of this discussion in answering our own question. The analogy is not perfect, for several important reasons, but it certainly helps motivate the problem of whether talents make for unfair advantages in a market or, for that matter, any other sort of economy. Moreover, the disanalogies may shed as much light as the analogies.

Economists define capital goods as man-made ones which usually increase the efficiency of, decrease the cost of, or otherwise further production. Capital goods can either be substituted more cheaply for other factors of production, i.e., labor and land, or they can be combined with labor and land to produce more goods than would be otherwise produced, or goods that otherwise would not be produced at all. One thing that is crucial about capital goods is that they typically satisfy consumer wants "indirectly" in "subsequent periods." That is, they are not immediately consumed but have a return in the future. Of course some goods can be consumed immediately or employed in a time-consuming productive process. The economist's classical example is grape-juice, which may be consumed when young or laid up to age and thereby improve into wine. But the usual case of a capital good is one that has a future use and is not the object of present consumption. Most talents have both of these features. They improve the productivity of their bearers in the provision of consumption goods, but are not themselves directly consumed. Of course, persons may derive pleasure from knowing they have a talent, but most often they "consume it" only indirectly through its products. And, in general, talents enable their bearers and others to produce the same output more efficiently, or to produce more of the output or to produce a new product that would be otherwise unattainable.

Like other factors of production, capital goods cost money to acquire and to maintain, and they get used up: they wear out or depreciate. Moreover, producing them requires shifting goods out of immediate consumption, in the expectation that once completed and installed the capital goods will increase production (or its value) sufficiently to repay those who sacrificed consumption in order to produce the capital goods. Some talents, such as the ability to play tennis, are just like capital goods in all these respects: they cost money to acquire (e.g., the cost of lessons), involve the sacrifice of current consumption (of leisure or fattening foods, for instance), and they can even wear out with a great deal of use (tennis elbow). Other talents, especially those we have called hereditary traits, are less like capital goods: they cost their bearers nothing to acquire, involve no sacrifice of consumption, and may never be depreciated by employment. Chuck Yeager's phenomenal eyesight at 70 years of age is a good example of a costless, undepreciating capital good. Other potential disanalogies involve talents that have become obsolete but, unlike obsolete capital goods, incur no storage or disposal charges. But this is a minor disanalogy at worst. Domestic talents may cost their bearer relatively little, while costing parents a great deal in time, money, and unpleasantness, like the skills of the musical prodigy acquired through years of expensive and annoying lessons and practice.

The biggest difference between capital goods and talents is that the latter are not transferable or tradable. At least for the moment, no market in alienable talents exists. What technological changes enabling organ transplants will do for the transferability of talents cannot yet be fathomed. On the other hand, it is because normal capital goods are transferable that they give us a handle on understanding the way in which talents provide advantages, for their transferability enables us to put an exchange value on capital goods which we can then in principle extend to talents. If we can assign a value to capital goods just because they are tradable, we may be able to assign a shadow price to untradable commodities with the same productive efficiency. Such a valuation might provide egalitarians a basis for charging the bearers of talents and compensating the bearers of disabilities.

How do we determine the value of a capital good? Once in place, a capital good provides a stream of returns in the form of increased output. Suppose a capital good wears out in t years but provides return (positive or negative) each year measured, say in dollars, $s_0, s_1, \ldots, s_t$. These are the annual payments to the capital good. If the annual interest rates are $i_1, i_2, \ldots, i_t$, then the value of the capital good at the beginning of production is given by:

$$\text{Present value} = s_0 + \frac{s_1}{1+i_1} + \frac{s_2}{(1+i_2)(1+i_1)} + \cdots \frac{s_t}{(1+i_t)\cdots(1+i_1)}$$

Thus, the present value of a capital good is the sum of the annual returns discounted by the interest rates in force in each period. If these returns are negative, then the capital good is a liability. Given the rate of interest for each period it is a relatively simple matter to decide whether to acquire a capital good offered at its present value: if the return on a security or a savings account bearing the prevailing interest rate is below the present value, then the capital good is a rational purchase. Otherwise, it is preferable simply to deposit the money in an interest-bearing account.

In an efficient economy the annual payments to a capital good, $s_0, s_1, \ldots s_t$, reflect its marginal productivity, that is, the amount of output that can be "credited" to it. Marginal productivity is a superficially tempting basis on which to erect a theory of justice in returns to capital, and to labor and factor inputs, for that matter. Since in a perfectly competitive market the wage rate and the rate of return to capital are equal to their marginal productivities, it might be held that the fair return to labor and capital equals their marginal productivities.

For they reflect the respective causal contributions to production, how much of the output each supplier of labor or capital or factor inputs is "responsible" for. This is, of course, a misunderstanding. Marginal productivity does not measure separable portions of the productive process. Each input is causally necessary: none is sufficient for any part of the output, and so none can claim *the* causal responsibility for a portion of it. Rather, marginal productivity serves as part of a solution to a bargaining problem. Each agent providing a factor has an incentive to withhold the factor for any payment below its marginal productivity, since other agents have an incentive to pay that price or forgo gains themselves. The annual payment to a capital good reflects its marginal productivity, and the present value represents the discount on this productivity over the future. The present value reflects the amount of money or other resources required in order to build or buy the capital good. The seller of capital will forgo gains at a lower price for the good, and the buyer of capital will forgo gains at a higher price. From the fact that present value measures increased productivity, it does not follow that the rightful owner of the capital good has a right to the proportion of increased annual output that it enables the user of capital to produce. For even if causal responsibility were a basis for compensation, it is no more *the* cause of this increased output than any of the other factors, including labor, which is required to produce anything.

The formula for present value gives the annual interest rate a crucial role in determining the present value of a capital good. As the interest rate increases, the present value of a future payment decreases, and vice versa. But why is this the case, and what determines the rate of interest anyway? The latter question is a much vexed one in economics. One thing all seem to agree on is that in the pure case, the interest rate reflects the degree to which individuals discount future payments as against current ones: given the choice between a dollar today and a dollar next week, the rational agent chooses a dollar today just because of the uncertainty of getting that dollar next week. To get him to take a payment of one dollar next week instead of today requires that we offer a premium, say, one dollar and fifty cents. The interest rate measures the premium required to get agents to abstain from current consumption. The amount of premium we must pay will vary from individual to individual, as a function of preferences for present and future consumption and, in a situation of uncertainty, for risks. Frugal people will postpone consumption at a lower rate of interest than spendthrifts, as will gamblers.

In addition to reflecting the preferences of those who choose between consuming present goods and investing them, the interest rate reflects the opportunities producers foresee to expand production by the use of capital goods. Given their beliefs about the demand for commodities, and the present value of the capital goods needed to produce them, they will borrow funds at interest, hoping to make a greater return than the rate at which they must pay off the loans. The supply of lendable funds, reflecting consumers' time preferences, and the demand for such funds, reflecting capitalists' expectations about consumers' preferences, together determine a market interest rate. For our purposes, what is crucial about the interest rate is its dependence on the tastes and preferences of economic agents for present and future goods. Exactly how individuals' preferences and expectations aggregate into a market interest rate need not concern us.

The return to the owner of a capital good reflects the preferences of agents who consume its output, but it also reflects the number of other owners of capital goods of the same type. Assuming a downward sloping demand curve for a type of capital good, the return to the owner is a negative, positive, or zero "economic rent." In the present context, this is a crucial notion because such "rents" are often viewed as an economically unwarranted return to capital. Sometimes the owner of a capital good will be paid more than the minimum needed to prevent its transfer to other users. This excess is its economic rent. Whether the owner of a capital good receives economic rent depends on the shape of the supply curve for that type of capital good.

In a perfectly competitive setting, in which there are indefinitely many suppliers of a commodity, say a capital good, no individual supplier can affect the market by withholding or increasing supply at the going price; and if he sells at a price above the market price, he will have no buyers, while if he sells below, he will simply forgo income he can gain without any further effort. If the price paid for a unit of the capital good is the long-run price under perfect competition, then a drop in the price will simply result in all suppliers shifting their capital goods to other uses, for otherwise they would be supplying their goods at a lower price than they could earn elsewhere. Here, the return to any one supplier is just the amount needed to prevent transfer to another use. There is no economic rent being paid.

Suppose, however, that the supply curve is perfectly vertical: the same amount of the good will be offered no matter what the price. In this case, the entire return to any individual supplier of the capital good is rent, for there is no alternative use for the capital good, and so no need to pay to keep it employed in its current use. The third and more usual alternative is a supply curve sloping upward to the right, intersecting the demand curve at some quantity of capital q and some price p that equilibrates supply and demand. At this point, every supplier of capital is paid p per unit of capital he supplies. But presumably most suppliers would have been willing to supply at a lower price. It is only the price demanded by the "last" supplier needed to equate demand and supply that actually equals p. But all suppliers receive p. Accordingly, each of the suppliers who would have been satisfied with less than p per unit of capital receives an economic rent equal to the difference between p and the price at which they would have supplied a unit of capital. The accompanying diagram reflects the three different alternatives. The shaded area represents the economic rent of capital when the supply is elastic.

For our purposes, it is important to note that economic rents have long been viewed as especially suitable targets for taxation. Sometimes this is defended on the ground that they are unearned: all but the "last" supplier would have supplied output at a lower price or, if the supply curve is vertical, at almost any price. Moreover, taxing rents will not deflect the allocation of capital from its most efficient uses, and so provides no disincentive to efficiency, unlike other taxes. As the diagram shows, there will be at least some economic rents in all but cases of perfect competition – in other words, in all actual cases, since perfect competition is nonexistent. Further, the phenomenon of economic rent is by no means restricted to capital, but will arise for labor or any other factor input with an upward-sloping supply curve.

III

Let us now apply this theoretical machinery to talents and disabilities. Most talents are arguably not objects of immediate consumption but, rather, may be employed to produce goods for immediate consumption. People do, of course,

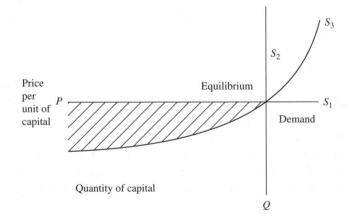

take pleasure in some of their talents, even when they are unexploited. And disabilities are sources of profound displeasure, because the disabled feel keenly the opportunities of which their disabilities have deprived them. But by and large, talents only provide fungible advantages when they are put to use in producing consumption goods, and disabilities fungible disadvantages because of the extra costs they impose on the production of immediately consumed goods. The increased production talent permits should, in principle, enable us to associate a present value with it, one which reflects the stream of annual payments in additional consumption goods it can secure as a result of its role in increasing production or lowering its cost.

Similarly, a disability will have a negative present value because of the annual extra costs it will impose on production or the reduced production it will cause. But determining the size of this present value requires, first, that we assign a fungible value to the annual returns and, second, that we determine an interest rate. In some cases, conventional, transferable capital goods might just conceivably provide a surrogate measure of present value. For example, the present value of the talent to calculate vast sums in one's head may be set equal to the present value of an electronic calculator with equivalent powers. Other talents, like Mozart's, are beyond price. But for our purposes, solving this problem of assigning present values is irrelevant. Rather, the point of the comparison of talents and disabilities to capital goods is to show in theory exactly what it is about a talent that makes it an economic advantage. It will also make evident how fiercely complicated it would be even in principle to secure the information needed to show that any type of biological difference was an advantage or a disadvantage for any given individual, let alone across the society as a whole.

A disclaimer: My skepticism about our ability to identify and price talents and disabilities is not meant to blind us to the great hardships and suffering borne by people afflicted by obvious disabilities, as well as unobvious ones. Rather, it serves as a caution against our hopes to be able to fine-tune social institutions that charge and compensate for talents and disabilities.

Whether a biological endowment is a capital asset will evidently depend on the tastes and preferences of agents for consumer goods and their time preferences, as well as the state of technology in the society. Moreover, unless the supply and demand for all goods in the market is equilibrated by *tâtonnement* in a Walrasian auction (where no bids are accepted until all prices offered perfectly balance supply and demand for every commodity) or by some other kind of recontracting that in fact never obtains in any real economy, then the return to the very same biological endowment will vary from person to person in a given economy. Let us see why.

Present value is a direct function of annual returns and an inverse function of the interest rate. What factors determine annual return? This figure is the amount of income from sales of a consumer good that can be imputed by a marginal productivity analysis to the capital good. This amount will vary with the demand for the consumer good, and also with the supply of capital goods of this type. The more scarce a capital good is, the more economic rent it can earn, and the greater is the portion of the return on sales for which its owner can bargain.

Thus, in a population of Polynesians with no interest in coconuts, an agile tree climber will have no capital advantage. And in one in which all crave coconuts and can climb equally well, he will have no advantage either, though in both cases his biological endowment remains unaltered. Suppose, further, that among Polynesians there is only a very low time preference. Most are perfectly happy to postpone immediate consumption of, say, sugar canes in order to use them as pruning hooks to reach the coconuts. Thus, the interest rate, measured in sugar cane, is so low as to make tree-climbing ability a relatively worthless talent, whether widespread or not. Under these circumstances it does not appear that the talent to climb trees bestows any economic advantage, and its present value is quite small at best. But things might be quite different. Suppose the coconut trees are too high for any method but climbing, and there is nothing to eat but coconuts. And suppose that time preference is quite high because, lacking coconuts today, the population will starve to death tomorrow. And suppose, finally, that everyone has acrophobia and only a handful have talent for climbing. Here, the present value of the talent becomes extremely high, and much of its return is in the form of economic rent as well.

There is another scenario worth contemplating for a moment. Suppose all can climb equally well but, while most have acrophobia, a few have acrophilia – the love of high places. Is the preference for low places a handicap? It clearly has a negative present value, since it requires individuals to pay more for coconuts in each period than they would if they lacked this preference. (Recall that both annual payments and present value can be negative amounts – costs instead of payments.) Is the preference for high places a talent? It certainly pays. What this case shows is that the economic effects of a talent or a disability may be exactly the same as those of a preference or taste. This is important, because some writers[10] have argued that individuals ought not to be compensated when their tastes make them do something or forbear from doing it, but should be when their disabilities do the same. If we cannot distinguish disabilities from tastes, say, the inability to climb from acrophobia, or distinguish those agents with the latter from those with the former, we will be unable to effectively compensate only for real disabilities and avoid compensating those whose preferences lead them to feign disability, or even those who sincerely believe themselves disabled. In a parallel fashion, we may end up charging someone for his acrophilia just because he climbs the tree with no more talent than an acrophobic. Of this problem about tastes and preferences, more later.

The real problem about identifying the present value of a talent is the un-evenness of preferences and opportunities characteristic of any real society changing in real time. Even the frictionless world of rational economic men needs a special device to find the market price of goods that equilibrates their supply and demand: the mechanisms of *tâtonnement* or recontracting. Once everyone's endowment of commodities and capital goods (talents and dis-abilities) is given and trade booms, many factors will determine the estimates people make of the present worth of a talent or disability.

Thus, if acrophobics are concentrated on the eastern part of the island, while agile tree climbers are in the west, the odd eastern tree climber will find that his talent earns him far more than a western climber. As news travels to the west of the demand for climbers in the east, and as news travels to the east of the surfeit of climbers in the west, the estimates of the present value of tree-climbing ability fall in the east and rise in the west. If enough tree climbers are found, the component of present value attributed to economic rent instead of allocative efficiency declines, and it may even reach a point at which tree climbing for coconuts becomes a sport no one is paid to perform. But even the real world of the Polynesian islands is not so simple, and it is doubtful that talents ever fetch their equilibrium prices. For this we would require perfect information about the present and future, perfect rationality, and an auctioneer who will allow no exchanges of goods and services until the parties to the exchange, by successive approximation, hit on a schedule of prices that equilibrates supply and demand in all markets.

Only at this point can the present value of a talent or disability be de-termined. Since in the real world information about the present is imperfect and about the future unavailable, the best we can do is make rough guesses about future payments and interest rates. Now, for economic purposes such guesses may suffice, and the better they are, the more money an individual can make. But return to our color-blind draft dodger in peacetime, or con-sider the well-coordinated seven-footer in a society that despises basketball. Are these persons differently endowed from those for whom color blindness and great height have substantial present value for themselves and/or others? Might not my gifts be substantially rewarded by the market if only I had found myself among people with different tastes and a different level of economic development, people who may even actually exist but, alas, elsewhere? Are not most of us in this situation? On the other hand, most of us have biological endowments that would be severe handicaps were the tastes of others very different from what they are. My excellent hearing would be a grave disability if I lived among adolescent aficionados of rock music. Unfortunately, except in the economist's ideal world, it is quite impossible to tally the payments and costs, and the local rates of interest, that would enable us to set a present value on our talents and disabilities. In the real world, most of our talents do not receive the return that they would earn in an economist's perfect one.

It is equally difficult for observers to distinguish many agents' talents from their tastes and preferences. Indeed, we may be unable to distinguish our own as well, because of fearful thinking that conflates a phobia with a disability and wishful thinking so strong that it compensates for lack of talent.

To sum up, then, biological endowments are difficult to identify even biologically, and differences among them are distributed in ways that are difficult to summarize in terms of normality and abnormality. The role of environmental factors in the fixing of biological endowments is so great that to call such endowments hereditary may be seriously misleading. Some of these environmental factors may be under the control of no agent, others under the control of parents, and others under an agent's own control. In the latter two cases, which environmental factors are brought to bear reflect both the resources and the preferences of agents or their parents. But whether an endowment is a talent or a disability depends on the preferences of those with whom the bearer comes in contact, and on how social institutions are arranged in the light of those preferences. The market value (or disvalue) of talents and disabilities will reflect the strength and distribution of preferences and tastes within the society: the more varied those tastes and the more complex and larger the society, the less likely it is that a talent will fetch its equilibrium market price. It is similar for a disability. Moreover, in the case of a talent, the amount of the advantage that comes in the form of economic rent unnecessary to insure allocative efficiency will be a function of the number of individuals who bear the talent.

Talents and disabilities can provide advantages and disadvantages in any society in which returns to factors of production reflect their marginal productivity, whether such societies are market economies or not (though in a nonmarket economy it may be physical marginal productivity that such returns reflect). This is because talents and disabilities, unlike transferable capital goods, cannot be expropriated. A society can nationalize a productive capital asset or an unproductive capital liability, and thus deprive its owner of the associated advantage or disadvantage, but it cannot do so to a talent or a disability. It can, of course, attempt to coerce a talent into employment, and subsidize a disability, but in both cases it faces the problem of incentives and disincentives: the undisabled will have incentives for pretending to be disabled to some degree, and the talented disincentives for employing their talents at optimal levels, unless, of course, the costs of employing a talent are nil or are outweighed by the benefits at any level of remuneration. Thus, we need not pay the bearer of a talent anything to exercise his talent, if he gets enough sheer pleasure from doing so, and whatever we pay him will be an economic rent. Of course, any one entrepreneur will have to outbid others to get the services of a talented agent, if there are other entrepreneurs. If all talents are like this, the incentive problem does not arise, and the advantages a talent provides can be eliminated by nationalizing all the returns to its employment. Similarly, we can imagine a society in which there is no incentive to pretend to a disability because there is no productive liability associated with it to subsidize. For example, a person confined to a wheelchair in a society in which even ambulatory persons prefer always to sit is not economically disadvantaged, and no one who can walk has an incentive to pretend otherwise.

IV

With this brief sketch of the problems surrounding the identification and measurement of talents and disabilities behind us, let us turn to the problem of whether or not these differences make for unfair advantages. It will be both convenient and illuminating to examine the question in the company of one who has studied it extensively, and come to definite conclusions on the matter: Ronald Dworkin.

Dworkin has argued at length that the most adequate conception of equality is that of equality of resources, and that equalizing resources requires compensation for disabilities and charges for talents. According to Dworkin, equalizing resources also requires a free market. Accordingly, the charges a political authority fixes for talents, and the compensations it makes for disabilities, in pursuit of a policy of equality must be consistent with market mechanisms. Before considering Dworkin's argument for these views, some stage-setting is required.

Why equality? As I noted at the outset, Dworkin explicitly abjures any argument for equality as a requirement for social organization. His stated aim is simply to determine which definition of equality is most adequate. But unless some justification for equalizing is offered, the entire exercise is too abstract to have any practical bearing. If each moral agent had a natural right to equality (of whatever kind is most adequate), the justification would be obvious. But if there are no natural rights, the claims of equality need some other foundation. The only one that seems to me remotely plausible is a contractarian one. We agree to mutually enforce equal treatment because that is the best each of us can do given the demands and the capacities of others. If this contractarian or bargaining view cannot be fleshed out, then there seems to me to be no compelling *argument* whatever for equal treatment, no reason to secure or sustain equality beyond our sheer noncognitive preference for it. In what follows, I shall assume that some such argument is forthcoming and that the grounds for equality have been secured. Without it the exercise is idle.

Equality of what? Dworkin argues that equality of welfare, satisfaction, utility, or similar matters of preference, is unsatisfactory for a variety of reasons, and that the most adequate kind of equality is equality of resources; and the test of equality of this kind is the absence of "envy," defined as the preference of any one agent for the resources of another. Among rational agents, no envy will be evinced when all resources are equally divided. This division, Dworkin holds, "presupposes an economic market of some form, mainly as an analytical device but also, to a certain extent, as an actual political institution."[11] A market of some form is needed, Dworkin writes, but the kind he requires is the unrealizable recontracting market of Walrasian *tâtonnement*, in which no trading is allowed until market-clearing equilibrium prices have been established for all resources. In a real market with finitely many traders who are relatively ignorant and capable of trading only with neighbors, envy can set in after the first free exchange.[12] In fact, even where envy does not set in, the result of real trade can be far from optimal, even on so weak a criterion of optimality as Pareto's. Dworkin avoids these problems of the real market. He notes, "I make all the assumptions about production and preferences made in Gerald Debreu's proof of the existence and Pareto optimality of a stable general equilibrium,[13] the most abstract result in neoclassical economic theory."

Now talents are clearly resources, and disabilities are the lack of them, so any equal distribution of resources must take account of them; it must either distribute the benefits they provide to all equally, or charge and compensate those who bear them exclusively. This is evidently because talents and abilities are not distributed in accordance with desert, nor are they earned, but they are distributed by brute luck. And distribution should not be a function of brute, unearned luck.

As we have noted, however, this is not in general true of talents. At least some talents and some disabilities are not the result of luck; they are earned: the talent for tennis, the brain damage that comes from willfully riding motorcycles without helmets, and other personal endowments. To the extent these endowments are earned, many will hold that they may not be subject to charges and compensation. Additionally other endowments, like the musical prodigy's gifts, have been paid for by parents (and not just in money, but in time, annoyance, and other inconveniences). Indeed, many of the most important talents individuals bear may have been earned by them or by others who have transferred these advantages in morally unquestionable ways. I dare say, the most significant advantageous endowment we have, intelligence, is thus secured not by random luck but by parental investment (and again, not just of money). Much the same may be said of intellectual disabilities. They are often the products of parents with enough resources to prevent them, and after a certain point are the responsibility of their own bearers. As for disabilities, we earn and deserve many of them, by drinking, smoking, overeating, and the like. To many, there seems little reason to suppose such disabilities require compensation of any kind.

Naturally, if we hold this view firmly, we will want to exclude at least some earned and deserved talents and disabilities from any scheme for equalization. But what if the biological endowments which provide the most widespread advantages and disadvantages are those which are earned. If this is indeed the case, then beyond succoring a small number of victims of brute bad luck, and charging an even smaller number of coordinated seven-footers who happen to live among people with a liking for basketball, the whole question of whether biological endowments make for unfair advantages will be academic.

So, focusing on the subset of talents and disabilities acquired through brute luck, the envy test for equality of resources requires that such endowments be equalized through charges and compensations. The aim of these charges, and especially of the compensations, is not to make up for having a talent or a disability. As Dworkin notes, no amount of money will compensate someone for blindness, i.e., make him the equal of a sighted person. But, according to Dworkin, given the chances of blindness striking through brute luck in spite of normal precautions, there is some amount of insurance each of us might pay a premium to receive in the event that we were struck blind. Our conception of talents and disabilities as capital assets and liabilities underwrites this. Insofar as a biological endowment is a productive factor, either asset or liability, the amount a rational agent might be expected to pay in premiums for insurance against disability, or the chance to own a talent, is the expected present value of the talent or disability, i.e., the product of its present value and the brute-luck probability of being afflicted by the talent or disability. Of course, these calculations must be made by a rational agent who does not already know

213

the value of the untransferable talents and/or disabilities he does or does not have. Otherwise, agents have incentives to hide talents and feign disabilities. A veil of ignorance is required to make the rational agent state his actual level of willingness to pay for protection from a disability or for the use of a talent.

But Dworkin holds that rational agents must at least know what talents they have. Otherwise, he writes,

> we have stipulated away too much of his personality to leave any intelligible base for speculation about his ambitions, even in a general or average way. The connection between talents and ambitions . . . is much closer than that between ambitions and handicaps – it is, for one thing, reciprocal – and much too close to permit that sort of counterfactual speculation.[14]

Without knowing both an agent's biological endowments and his ambitions, i.e., his tastes and preferences, we cannot tell how much a talent or disability is "worth" to him. Moreover, as we have seen, a preference may have exactly the same consequences for an agent as a talent or disability. Recall the effect of acrophobia and acrophilia among our Polynesian islanders above. This raises a serious problem for a scheme like Dworkin's.

According to Dworkin, we need not treat tastes and preferences on the same footing as biological endowments. Though some preferences are debilitating enough that a rational agent might take insurance against them, in general preferences ought not be subject to compensation, or for that matter to charges, and this for two reasons. First, "we cannot state . . . what equality in the distribution of tastes and preferences would be."[15] Second, preferences, unlike mental or physical powers (or their absence), are not parts of an agent's "circumstances." Rather, they are parts of his definition of "what a successful life would be like";[16] they are "features of personality";[17] they are assigned to the person. Accordingly, equality of resources need not take into account differences in ambition, even though they may have the same effects as differences in endowments. This is troubling because, as we have seen, preferences can have exactly the same effects as talents and disabilities.

Moreover, the two differences Dworkin cites do not suffice to justify such different treatment. First, it seems no easier to determine equality among bundles of talents and disabilities than among sets of preferences. Is near-sightedness plus manual dexterity equal in present value to mild deafness plus foot-speed? We can answer this question only if we know more about the market for these bundles than we can in practice ever know. But is the problem of equating preferences any more difficult than this? The envy test might help us determine whether these two bundles were equal in preferability *modulo* our preferences, though I for one cannot tell how it would come out. But such a test might also enable us to equate different bundles of preferences *modulo* opportunities to satisfy them. If not, the difficulty of stating what equality of preferences would be is no reason to treat them differently from talents and disabilities.

The second reason for different treatment of preferences and talents is that the latter are parts of the person, while the former are parts of his circumstances. I confess I do not see what difference this makes, to the extent that I understand it. For one thing, powers and abilities are not parts of the agent's environment, which he can change just by moving away or outwaiting their change. It is true that the powers I was born with, or endowed with by others, are parts of my circumstances in the respect that I did not choose them, may not deserve them, am not responsible for them, and can alter them only by great effort. But exactly the same is true of my preferences, my ambitions, and my definition of what a successful life would be like. For they are determined by factors beyond my control as much as my endowments are.[18]

Moreover, whether an agent's biological endowment is a talent or a disability, we have seen, depends crucially on his own preferences and on those of others among whom he finds himself. To the draft dodger color blindness is a godsend. To the would-be jockey great height is disability. So it goes for every biological endowment. To charge for talents and compensate for disabilities is *ipso facto* to do the same for preferences. For talents are complexes of endowments and preferences – preferences of the bearer of the endowment and of others as well.

This makes a problem for Dworkin's approach, for he argues that justice requires us to equalize for talents but not for preferences. Our test should be sensitive to ambition but not to resources. "On the one hand we must on pain of violating equality, allow the distribution of resources at any particular moment to be . . . ambition-sensitive. . . . But on the other hand, we must not allow the distribution of resources at any moment to be endowment-sensitive, that is, to be affected by differences in ability of the sort that produce income differences in a laissez-faire economy among people with the same ambitions."[19] But to allow the distribution of resources to be ambition-sensitive makes them resource-sensitive, since an endowment is a talent or a disability only *modulo* an ambition, a preference. If we deprive someone of the gains of a certain ability, we are depriving him of gains due to his ambitions and others' preferences, for those are what turn his endowments into abilities and talents. To charge or compensate on the basis of brute biological endowment, with no eye to ambitions, on the grounds that these are not subjects for equalization is an impossibility.

On Dworkin's view, we have no idea of what equality in the distribution of tastes and preferences would be. If so, we have no idea of what it means for people to have the same ambitions, unless it means simply for everyone to have exactly the same preference ranking over all possible bundles of commodities, occupations, leisure vs. working hours, and so forth. This sort of equality of ambition would, together with a distribution of biological endowments, fix talents and abilities in a way that exactly mirror's the distribution of the endowments. Then, provided we can solve the problem of the incentive to hide talents and magnify disabilities, we could charge and compensate to equalize the outputs of production. But the slightest differences among people in preferences destroys this result. Given anything other than this abstract possibility, no distribution can be endowment-insensitive while at the same time being ambition-sensitive.

One difference that distinguishes at least some of both the most well-off and the least well-off from the median in any society is divergent preferences and antipathies towards risk. Some individuals have a taste for high-risk gambles with large stakes and vast payoffs, and others do not. A majority of the members of the former class are likely to end up poor and a minority quite well-off. If the quantum of risk preference or aversion an agent has is a matter of brute luck, and untransferable luck at that, then individuals with the same endowments, say a head for figures, may end up quite differently advantaged as a result of preferences only. The risk-preferring lightning calculator will probably end up at a wealth level quite far from the median, one way or the other, through the undeserved, unearned, brute-luck draw of a high risk tolerance. Ambition in this case has had the same effect as a talent or a disability. For it will turn an endowment into a disability in all but a minority of cases – the lucky winners who break the bank at Monte Carlo.

This, of course, may be taken as an argument that we should equalize for ambitions and preferences, as well as for talents and disabilities. Indeed, Dworkin should in all consistency so treat it. Aside from the conceptual problems with doing so that Dworkin itemizes in "Equality of Welfare," there may be moral objections as well to the notion that someone should get less or more of a scarce commodity solely as a function of his desire for it. Additionally, this argument suggests that equalizing for talents and disabilities is no easier than equalizing for preferences. We can show this in another way by following Dworkin's ideal means of implementing the insurance mechanism for protecting ourselves against brute-luck disabilities and paying for talents.

The premiums we should pay, and the insurance benefits some of us would receive, can be extracted in the form of a redistributive income tax. Such a tax is ambition-sensitive because it allows those with the ambition to be rich to amass wealth by the dint of extra labor and greater savings, while enabling those whose ambition is leisure and immediate consumption to acquire them at the expense of future wealth. By taxing income instead of wealth it leaves ambitions alone, and by taxing it in a graduated way it may enable us to equalize income between the talented, the average, and the disabled. Dworkin recognizes the difficulty of this proposal:

> [T]he appeal of a tax depends on our ability to fix rates of taxation that will make th[e] compromise [between talent and ambition] accurately. It might be helpful, in that aim, if we were able to . . . identif[y] in any person's wealth . . . the component traceable to differential talents as distinguished from differential ambitions. We might then try to devise a tax that would recapture, for redistribution, just this component.[20]

I have argued that this cannot be done, because biological endowments and ambitions are equal determinants of talent and disability. Dworkin comes close to recognizing this point. He continues the passage just quoted:

> But we cannot hope to identify such a component, even given perfect information about people's personalities. For we will be thwarted by the reciprocal influence that talents and ambitions exercise on each other. Talents are nurtured and developed, not discovered full-blown, and people choose which talents to develop in response to their beliefs about what sort of person it is best to be.[21]

The situation is even more complicated than this passage suggests. No biological endowment is a talent or a disability except *modulo* ambitions. And the genetic luck that makes for "unfair" differences in talent[22] also produces differences in ambitions with exactly the same effects on prosperity. Shall we tax both or none?

Because talents cannot be distinguished from ambitions for purposes of determining the level of redistributive taxation, Dworkin suggests the following idealized mechanism for calculating rates of taxation that will simulate an insurance system against unfair advantages. He begins with the assumption that agents know what talents they have but do not know either the demand for or the supply of any talent in the society – he expresses this point in terms of agents' ignorance of the economic rents of their talents. I have already argued that this level of ignorance precludes anyone's knowing whether he has a talent or disability at all, since whether a biological endowment is a talent or a disability depends on the bearer's preferences and those of the other agents in the society. To know that you have a talent, as opposed to an endowment, requires knowledge of your own preferences and those of other agents in the society (and among rational individuals this knowledge induces an immediate motive to understate or overstate the minimum income level he will insure against).

We shall see how this consequence blocks Dworkin's proposal. Given this knowledge of individual talents and ambitions by each agent, Dworkin introduces a computer into which data are stored about

> tastes, ambitions, talents, and attitudes toward risk ... as well as information about the raw materials and technology. ... It then predicts not only the results of the [Walrasian *tâtonnement* type] auction but also the projected income structure – the number of people earning each level of income – that will follow the auction once production and trade begin, on the assumption that there will be no income tax. Now the computer is asked a further ... question. Assume each [agent] ... knows the projected income structure but is ignorant of the computer's data base ... and is therefore ... uncertain what income level his own talents would permit him to occupy. ... How much ... insurance [against being locked into an unacceptably low level of income] would the [rational agent] ... buy, ... and at what cost?[23]

The answer to this question provides the minimum level of positive taxation for talents, via obligatory premium payments, and negative taxation for disabilities, via insurance payments, that equality of resources sanctions. Dworkin suggests that the rational choice will be a minimax strategy: agents will not insure against extremely high minimum incomes, because the premiums would be so high that the "fortunate ones" who through brute luck bore the talents to earn such incomes would have to devote most of their gains to paying the premiums. They would therefore be "slaves to their talents," unable to afford to take less remunerative jobs or produce at lower levels of output, because they could not then pay their premiums. Those who, on the other hand, were unlucky in the lottery for talents would not be burdened by a staggering debt they could never hope to discharge. But for lower levels of talent it becomes more and more rational to purchase insurance, because its cost will decline more rapidly than the minimum earning level insured. So, there is a graduated level of payments that rational individuals will be willing to make for insurance against having less than a certain earning power as a result of having less than a certain level of talent. Taxes pitched at the level of these payments come as close as we reasonably can to equalizing resources, given inequalities in talents.

It is worth pausing for a moment to ask what is wrong with slavery of the talented to the taxing authority. The answer that springs immediately to mind reflects the intuition that making them work harder than their preferences dictate violates their rights: people have a right not to exercise their talents to the fullest. But this right is empty if they must do so willy-nilly in order to meet their financial obligations to the state. Unless rights at least sometimes "trump" equality, the enslavement of the talented is required by equality of resources. And anything less than compensation equal to marginal productivity looks like at least some degree of slavery. As Nozick says, "Taxation of earnings from labor is on a par with forced labor."[24]

Dworkin's prescription means, of course, that some individuals will find themselves having paid taxes equal to a small premium for insurance while being endowed with very great talents. The trouble with this outcome is that it fails the envy test, as Dworkin notes. An agent not so advantaged will envy people who are taxed below a level that equalizes income, even though the envious agent has the level of talent or its financial equivalent that the rational agent would bargain for in an ideal situation. To cite Dworkin's example: Claude may point to the movie star and argue that his after-tax income violates equality of resources and is therefore envied by Claude, even though the level of taxation is the progressive one the computer found for the whole society on average, given data about talents, ambitions, tastes, risk preferences, and the sort.

> Claude may truly say that the difference between him and the movie star does not reflect any differences in tastes or ambitions or theories of the good, and so does not in itself implicate our first, ambition-sensitive requirement of equality in wage structure. . . . [Nevertheless,] Claude needs some argument in favor of [a] change . . . which is independent of his own relative position. It is not enough for him to point to people, even those of the same ambitions . . . as himself, who do better as things are.[25]

The closest we can approach equality of resources leaves some envious of others, because it does not equalize incomes entirely. We cannot more closely approach equality, Dworkin speculates, because of "the wholesale effects of any scheme of distribution or redistribution on the lives which almost everyone in the community will want and be permitted to lead. . . . Equality of resources is a complex ideal . . . an indeterminate ideal that accepts, within a certain range, a variety of different distributions."[26]

But now suppose we substitute for Claude's complaint that the rich movie star has his ambitions and the brute luck of movie-star talents. Now the complaint is that while both have the same endowments, the movie star has great wealth and Claude does not, because they have different ambitions and tastes, different attitudes towards risk. Claude envies the movie star his income, and complains that brute luck dealt him short in ambitions and tastes of a kind which would have given him an equal chance at the movie star's income. By parity of reasoning to Dworkin's treatment of identical ambitions and differential brute-luck talents, there ought to be an insurance-based tax scheme to equalize the effects of ambitions and tastes. Why, after all, should Claude suffer because his tastes and other people's tastes do not mesh in a way that provides him with a large income? No more reason than that he should suffer for his endowments not meshing with his own and others' preferences. And no less.

Unless, of course, ambitions and tastes are essentially different from talents and disabilities. And this, I think, is a crucial issue for those who claim we should compensate and charge for talents and disabilities. For they demur at the prospect of doing so for preferences and tastes. As Dworkin says, the latter define the person. What this means, I think, is that we tend to be determinists about talents and disabilities, and libertarians about tastes and preferences. People are not responsible for the former: they are a matter of brute luck – we could not have done otherwise than end up with the talents and disabilities we have. So they should not enter into calculations about desert, praise, blame, reward, punishment, and so forth. The same is not said for preferences and tastes. The Kantian idea that motives are what morally count stands behind our attitude towards preferences. For motives are just combinations of wants and beliefs, and beliefs can only be true or false, not moral or immoral.

Is this double standard of moral responsibility for preferences and not for talents and disabilities consistent? It cannot be to the extent that talents themselves are composites of endowments and preferences. On the other hand, preferences are themselves contingent on biological endowments, as Dworkin recognizes. (My point here is different from John Roemer's argument[27] that preferences are based on talents, and so must be taken into account. To the extent that differences in preference can be determined by differences in talents and disabilities, any compensation or charge for them will simply be a side effect of charges and compensations due directly to the talents and disabilities that underlie them.)

I think there is no fundamental *moral* difference between tastes and preferences, and talents and disabilities. Either we are responsible equally for both, or for neither in the metaphysically interesting sense of responsibility, that is, the sense in which responsibility requires a will free from determining causes. Of course, many moral philosophers are compatiblists and want to consider moral questions on the assumption that the question of free will is neutral with respect to many of them. Though I do not think this is ultimately possible, let us proceed a bit further on the assumption that it is. Instead of viewing responsibility vs. determination as an all-or-nothing matter, they consider that there are a range of preferences, from those for which we have no responsibility to those which are entirely up to us. Now, as I have argued, there is in this sense similarly a range of talents and disabilities from those wholly hereditary to those we choose to acquire (at various costs) ourselves. Even if Dworkin does not expect that the latter sort of talents and disabilities should be taxed or subsidized, he should still be willing to countenance (positive or negative) taxation of those preferences over which the bearer never has any control, but which advantage him or disadvantage him.

For some egalitarians this might be a tempting suggestion: we should approach as closely as possible to a scheme for compensating and charging agents to equalize for inequalities in advantageous and disadvantageous preferences. This course is not open to Dworkin, or to anyone who has thought about the problem of equality as thoroughly as he has. For, as Dworkin has shown at length, attempting to equalize for the satisfaction of preferences is a hopelessly incoherent idea. For all their impact on distributions, preferences are utterly intractable objects for measurement, and still less for counterbalancing against one another intra- or interpersonally.

The upshot is that any redistributive scheme will have to equalize for talents and preferences or for neither. Moreover, any very fine-grained scheme for redistributing will make for very heavy demands on their distributor's information about preferences. And since preferences are easy to hide, the redistributor will face serious problems of "incentive compatibility," problems of inducing agents to reveal their actual preferences, not to mention their endowments, while imposing a cost on them for doing so. And in the end, because talents and disabilities cannot be separated from preferences and ambitions, Dworkin's scheme will simply turn out to be an income tax which is not justified by the demands of equality of resources alone. As such, it will have to find a different justification, one which defends itself against Nozick's charge that it is tantamount to forced labor.

Even leaving these problems aside, Dworkin must deal with the problems facing anyone who hinges policy prescriptions on arguments from abstract general equilibrium considerations that are impossible to realize in nature. Considering the problems associated with drawing prescriptions about antitrust policy, for instance, from this theory, the tasks facing anyone ready to implement Dworkin's scheme will be daunting.

V

There are people with grave disabilities, and there are others with wonderful talents. Then there are the rest of us. We are doubtless all better off because of the benevolence of the latter to the former, and we would all be still better off if there were greater benevolence. The general obligation of those vastly enriched by their talents to those grievously deprived by their disabilities may well be agreed to by almost all, regardless of what moral theory they embrace and regardless of whether any moral theory is the right one. What seems to be more crucial for the character of society is whether those of us in between the most disabled and the most talented have a claim or an obligation in the light of our talents and disabilities. That talents and disabilities provide advantages and disadvantages in a market economy is palpable, especially on the analysis I have offered in terms of capital goods. Whether they are unfair is another question.

In considering whether or not these advantages and disadvantages are unfair in a market economy, we should bear in mind that if they are, they will probably be unfair in any other economic system, including the *nomenklatura* of a centrally planned one. For talents are far more difficult to nationalize than observable and transferable capital goods. Even if we conclude that talents and disabilities make for unfairness in a market economy, the free marketeer's breast beating must be tempered by the realization that things are no less unfair under other economic dispensations. This is not an argument for complacency about capitalism, just a reminder that the original question's restriction to market economies may not be very significant, even if we decide that differences in biological endowments are unfair in a market economy.

But there is one thing about the market that suggests that, by and large, differences in talents and disabilities are not unfair. The market institutionalizes the idea that people have an effectively protected right to what they earn. For absent this assurance, there is no reason to accept the market as a mutually advantageous arrangement. If most differences in talents and disabilities are earned, as opposed to unearned, then agents have the right to the advantages that these earned differences provide. And if they have the right to these advantages, they could hardly be deemed unfair in a market economy.

Of course, the question remains open whether people have a right to what they earn. Libertarians like Nozick certainly embrace this principle. And so do some of their opponents; Rawls, for example, argues that the returns to talents need to be redistributed just because the talents are unearned.[28] The principle is certainly not unarguable: Dworkin rejects it because equality of resources requires continuous redistribution via an income tax. His rejection is reflected in his treatment of a principle of equality of opportunity as opposed to equality of resources. He calls this the starting-gate theory of fairness:

> [I]f people start in the same circumstances, and do not cheat or steal from one another, then it is fair that people keep what they gain through their own skill. . . .
>
> The starting-gate theory holds that justice requires equal initial resources. But it also holds that justice requires laissez-faire thereafter. . . .[29]

Dworkin rejects this view on the ground that it is an inconsistent theory of justice requiring equal distribution at the outset, and permitting unequal distributions thereafter. He asks, if unequal distributions are permitted later, why not initially? And contrariwise, if equality is required at the outset, why not thereafter? "[T]he starting-gate theory, that . . . [agents] should start off equal in resources but grow prosperous or impoverished through their own efforts thereafter, is an indefensible combination of very different theories of justice."[30]

Dworkin's theory, by contrast, requires continual redistribution at least of income, if not of wealth. Equality of resources must hold everywhere and always if it holds at the initial position. I have said that Dworkin rejects the principle that earned advantages are fair, but this is not strictly correct. Rather, he rightly holds it to be incompatible with a principle of equality of resources, but he nowhere fully endorses this latter principle as more than the most coherent version of a principle of equality. Dworkin, we must remember, stops short of asserting the moral claims of equality.

However, rejecting this principle may be incompatible with the laissez-faire attitude Dworkin manifests with respect to preferences. These Dworkin will not factor into redistribution because they are part of the person, instead of his circumstances: they are truly "his" and, in a sense, they are "deserved" and must be respected. The advantages, if any, that they bestow Dworkin does not think unfair. Why? What is so special about preferences in contrast to talents? Could it be that they are, by and large, earned? If this is what stands behind Dworkin's respect for them, then he does after all endorse the principle that earned advantages are fair ones. The alternative to embracing this principle is equalizing for preferences as well as talents and disabilities, as I have tried to show above.

Suppose, therefore, that we accept the principle that earned advantages are not unfair. Now add to it the claim that most talents and disabilities are earned, or at least that the most economically significant ones are: the ones that make the most difference to how much education we get, what kinds of work habits we can maintain, how well we can get on with others, and how healthy we can keep ourselves. Together, this principle and this factual claim suggest that most biological endowments do not generate unfair advantages in a market economy. If the principle that earned advantages are fair is accepted, and if the factual claim that most talents and abilities are earned is correct, then the question of whether we should compensate unearned disabilities and charge unearned talents becomes an academic one. For there are relatively few of either, and the amount of redistribution needed in a wealthy society to succor the former will go unnoticed when expropriated from the latter. The result for general distribution of income will be little different from the disparities which the market always generates. And the charge that these inequalities of income are unfair because they are due to a morally arbitrary distribution of biological endowments will not be sustainable.

Behind this conclusion stand several different considerations for which I have argued in this essay. First, the notion of a biological endowment is a complex one, covering widely different kinds of talents and disabilities. Second, insofar as we can treat talents and disabilities as capital assets or liabilities, preferences come centrally into their valuation – both those of the endowment owners and of everyone else. And third, equalizing for resources by compensating for differences in talents and disabilities requires treading on preferences as well. Those who deny the permissibility of equalizing for welfare because of difficulties in or the unpermissibility of weighing interpersonal differences in preference cannot, *mutatis mutandis*, advocate compensation for differences in biological endowments either.

NOTES

Revised and reprinted with permission from *Social Philosophy and Policy*, vol. 5 (1987), pp. 1–31.

I must thank David Glidden, Onora O'Neill, and Ellen Paul for improvements on earlier versions of this chapter.

1. John Rawls, *A Theory of Justice* (Cambridge: Harvard University Press, 1971), pp. 72–74.
2. Robert Nozick, *Anarchy, State, and Utopia* (New York: Basic Books, Inc., 1974), pp. 224–227.
3. Ronald Dworkin, "What Is Equality? Part 1: Equality of Welfare," *Philosophy and Public Affairs*, vol. 10 (Summer 1981), pp. 185–246, and "What Is Equality? Part 2: Equality of Resources," ibid. (Fall 1981), pp. 283–345.
4. Nozick, *Anarchy*, p. 233.
5. See ibid., pp. 213–227.
6. For a discussion of this matter, see Amartya Sen, "The Moral Standing of the Market," *Social Philosophy and Policy*, vol. 2 (Spring 1985), pp. 1–19; and Allan Gibbard, "What's Morally Special about Free Exchange?" ibid., pp. 20–28.

7. As noted below in Section V, Dworkin for one would reject this line of reasoning. See Dworkin, "Equality of Resources," pp. 308–310.
8. For a discussion of this point see E. Sober, "Evolution, Population Thinking, and Essentialism," *Philosophy of Science*, vol. 47 (1980), pp. 350–383.
9. Dworkin, "Equality of Resources," p. 332.
10. See ibid.
11. Ibid., p. 284.
12. See Hal Varian, "Dworkin on Equality of Resources," *Economics and Philosophy*, vol. 1 (1985), pp. 110–127.
13. Dworkin, "Equality of Resources," p. 287 n. 2.
14. Ibid., p. 316.
15. Ibid., p. 302.
16. Ibid., p. 303.
17. Ibid., p. 304.
18. Nozick, *Anarchy*, p. 214. He makes this same point eloquently for quite a different purpose.
19. Dworkin, "Equality of Resources," p. 311.
20. Ibid., p. 313.
21. Ibid.
22. Ibid., p. 314.
23. Ibid., pp. 316–317.
24. Nozick, *Anarchy*, p. 169.
25. Dworkin, "Equality of Resources," p. 330.
26. Ibid., pp. 333–334.
27. John Roemer, "Equality of Talent," *Economics and Philosophy*, vol. I (1985), pp. 151–188.
28. Rawls, *a Theory of Justice*, pp. 72–74.
29. Dworkin, "Equality of Resources," p. 309.
30. Ibid., p. 310.

11

Research Tactics and Economic Strategies

The Case of the Human Genome Project

In the Museum of Science and Technology in San Jose, California, there is a double helix of telephone books stacked in two staggered spirals from the floor to the ceiling twenty five feet above. The books are said to represent the current state of our knowledge of the eukaryotic genome: the primary sequences of DNA polynucleotides for the gene products which have been discovered so far in the twenty years since cloning and sequencing the genome became possible.

1   THE ALLEGORY OF THE PHONE BOOKS

In order to grasp what is problematical about the human genome project (HGP), I want you to hold on to this image of a stack of phone books, or rather two stacks, helical in shape. Imagine each of the phone books to be about the size of the Manhattan white pages, and that the two stacks of phone books reach up a mile or so into the sky. Assume that the books are well glued together, and that there are no gusts of wind strong enough to blow the towers down. The next thing you are to imagine is that there are no names in these phone books, or on their covers. Only numbers. We do know that each phone number is seven digits long and we know the numbers have been assigned to names listed alphabetically, but without the names we can't tell to whom a number belongs. Moreover, the numbers are not printed in columns down the pages that will enable you to tell where one phone number ends and the next begins. Instead of being printed in columns down the page, the numbers begin at the top left and fill up the page like print without any punctuation between them. They are grouped within area codes, of course, and we can tell when one area code list stops and another begins, but we don't know the area codes, still less what geographical area they cover.

Sounds like a set of phone books that would be pretty difficult to use, doesn't it? Well, let's make them harder to use. Of course none of the individual phone books have names or any other identifying features on their covers. In fact, the books don't have covers, and what's more the binding of each directory was removed before the stack was constructed, and a random number of successive pages of adjacent phone books were rebound together. This rebinding maintains the order of pages, but it means that each volume begins and ends somewhere within each directory, and there is no indication of where these beginnings and endings are.

Can we make our mile and a half stack of phone books even harder to use? Sure. Imagine that somewhere between 90 and 95% of all the phone numbers in all the phone books have been disconnected, or have never even been assigned to customers. And of course we don't know which ones they are. These unused numbers look just like sequences of assigned phone numbers, and they even have area-code punctuation, though there is no geographic area assigned to these area codes. Remember, we can't tell which area codes represent a real area and which do not. We do know that between 5 and 10% of the numbers are in area codes which have been assigned, and that within these assigned area codes there are long lists of phone numbers of real phone company customers.

Although we don't know which are the area codes that are real, nor where they are in the directories, we do know some interesting things about these area code phone number listings. First, sometimes area codes and their phone numbers are repeated one or more times rather close together in a single volume, sometimes they are repeated in distant volumes in the stack: second, sometimes there are sequences of phone numbers which are very similar in digits to the numbers in a real area code, but their area codes are unassigned and all the phone numbers are unused. Even within almost all of the real area codes, the lists of assigned phone numbers are interrupted by long sequences of digits which when grouped into phone numbers are unassigned; sometimes within a real area code there are several of these sequences, longer than the sequences of assigned phone numbers within the area code.

Perhaps you are tiring of all the bizarre details of this idea of telephone numbers impossible to read. So, I will stop adding detail to our picture. But don't let go of the picture. Imagine that someone, a numerologist, say, now proposes to you that for three billion dollars of the U.S. government's money, he will put together a team that will transcribe all the digits in the mile and a half double stack of phone books into a computer. It's not the phone numbers he offers to transcribe, just the digits – one after the other – unsegmented into the phone numbers. The numerologist promises to make the list of digits available to anyone who asks for it, free. Assume further that three billion dollars is the cost of copying out the numbers with no hidden profit for the numerologist.

I suppose one's first reaction to such a proposal would be to thank the numerologist for his offer, but to decline it on the grounds that the list of digits is of no immediate use to anyone, even if we were going to have a very large party and wanted to invite everyone who had a phone number. But it is in our nature as philosophers to wonder, so we ask our numerologist, "what's in it for you, why would you do this transcription at cost, without any profit?"

Imagine our numerologist is candid and comes clean as follows: He is not really a numerologist, but rather represents a relatively large number of privately held direct-sales companies, each of which has a potentially very useful product, which it can only sell over the telephone. The companies know that there are enough potential customers out there to go around so that each of their shareholders can become rich through the sale of the very useful product they can manufacture, if only they had the phone numbers of the customers. So, our numerologist direct-sales marketing representative says that if the companies he represents had all the digits, they could sell the products, make every customer better off, and become rich themselves.

There are two responses one should make to the numerologist's admission: the first is, even if you can segment the three billion digits into phone numbers, why should the government pay for the phone lists, if you and the consumers are the ones to profit? More important, surely putting all those digits from all those books into a large computer file, without being able to tell the meaningless ones from the meaningful ones, just for starters, is not the best way to get in touch with potential customers.

Consider the first question, why should the government pay for the phone list? The answer given on behalf of the direct-sales marketing companies is that their products are guaranteed to help people stay healthier and live longer. But in that case, if they think putting all these phone numbers in a computer memory is so valuable, why don't they arrange to fund the project themselves, and reap the rewards by selling the product? The answer we get is that their capital is tied up in even more valuable investments, and besides, there is a free-rider problem. If any of them get together to fund the transcription, the disks on which the transcription is recorded could easily fall into the hands of other members of the trade association without these firms paying. So, we might reply, what's it to us? Why should the government get you out of this predicament? If it does, will you cut in the government on your profits from using the phone numbers? Oh no, comes the reply. That would be a disincentive to developing new products to sell to people on the phone list.

But wait a minute, let's turn to the second question. What's the use of a transcription of all these digits? Aren't there far better ways to get the names of customers with telephones? In fact, is this any way of getting the names of customers at all? Well, comes the response, what if developing the technology to transcribe all these phone numbers will also enable us to identify the real area codes, to segment the digits into meaningful phone numbers, and to begin to tell which ones are actually in use? That might justify some investment in transcribing the phone numbers. But unfortunately our numerologist can make no such assurance. At most he can promise that once we have the total list of real area codes, that 5 to 10% of the list of digits will be real phone customers. Are there ways of identifying these area codes, we ask. Certainly, says our interlocutor. Well, what are you waiting for, go and find them. When you have done so, you may or may not have any use for the transcription of all the digits. But until then, you would be wasting your time, or someone else's, along with a lot of money, government or private, to transcribe all these digits, including the ones in the unassigned area codes and the numbers with no customers.

This allegory, I suggest, pretty well matches the biochemical facts about the human genome, the molecular structure of the nucleic acids that compose it, the prospective payoff to sequencing the three billion base pairs of the human genome, and the policy advocated by the proponents of the HGP. Here is a brief explanation of my simile between the human genome and the mile-high stack of phone books: the human genome contains about three billion base pairs of purine and pyrimidine nucleic acids. These are the digits in our "phone numbers." It is hoped that the average cost of providing the whole sequence can be brought down to one dollar per base. Three of these bases constitute a phone number with three digits – a codon of three nucleic acid bases. And 90 to 95% of these sequences code for DNA that has no role in gene transcription – this is the so-called junk DNA.[1] They compose phone numbers that are not in use. Of the 5 to 10% of the nucleotide sequences that do code for proteins, we know little that distinguishes them from the non-coding "junk." These sequences of codons that do code for gene product are the phone numbers in a real area codes. Unfortunately we can't tell them from area codes that contain only "phone numbers" not in use. In our story we didn't know how many area codes there are, where they are in the phone books, how often they are reprinted, and how many corrupt sequences of meaningless phone numbers they contain.

Similarly, we do not know whether there are 50,000 genes or 100,000. We don't know their nucleotide size, the number of their copies, their locations, the number of stretches of meaningless numbers – the introns within each of them. The DNA for gene products is dispersed throughout the non-coding part of the genome. Within regions that code for gene products, there are long stretches of polynucleotides that, like nonsense or junk DNA, code for nothing, and whose messenger RNA sequences are deleted before protein synthesis. Some gene sequences are repeated two or more times throughout the genome, on the same or different chromosomes, as are some stretches of junk DNA. Sometimes the same strand of DNA can code for two different gene-products depending on codon-punctuation. Although we know what the start and stop codons are, we don't know how to segment the sequence of bases to tell either where a gene for some product begins, or which of the possible triplets of successive bases is the reading frame for the product – we don't know which among the digits in a sequence is the first digit of a phone number, even though we know that each phone number is three digits – three nucleotides long.

2   STRAW ALLEGORY?

How can exponents of the HGP defend its scientific integrity? One thing they cannot do is to respond to the allegory with a denial that the HGP really contemplates sequencing three billion base pairs of mainly nonsense DNA as one of its primary missions. Such a reaction would of course grant the aptness of the allegory and admit the pointlessness of the project of completing the sequence. If it were true that sequencing the entire genome were not integral to the original or current aim of the HGP its proponents would of course be guilty of seriously misleading the governments which support the project. But the fact is, sequencing the whole human genome has been and still is the goal of the HGP.

Writing in the fall of 1993 in *Gene* the original director of the National Center for Human Genome Research, James Watson, described the HGP as follows:

> Its mission was not only to make much higher resolution genetic maps but also to assemble all the human DNA as overlapping cloned fragments running the entire length of all the human chromosomes. In their turn, these DNA pieces were to be sequenced and their respective genes revealed. Upon completion of the Human Genome Project we would then know how many human genes exist

and so whether the then estimated 100,000 number was either too low or too high. (Watson, 1993, p. 309)

But making physical maps at varying scales of the human genome, and sequencing all the DNA will not answer these important questions about how many genes there are.

In October of 1993 Francis Collins, who replaced James Watson as director of the National Center for Human Genome Research, and David Galas, associate director of the Department of Energy Office of Health and Environmental Research, wrote:

> Although there is still debate about the need to sequence the entire genome, it is now more widely recognized that the DNA sequence will reveal a wealth of biological information that could not be obtained in other ways. The sequence so far obtained from model organisms has demonstrated the existence of a large number of genes not previously suspected. . . . Comparative sequence analysis has also confirmed the high degree of homology between genes across species. It is clear that sequence information represents a rich source for future investigation. Thus the Human Genome Project must continue to pursue its original goal, namely, to obtain the complete human DNA sequence. (Collins and Galas, 1993, p. 46)

This passage is quite revealing for what it omits. It is certainly true that knowing the entire sequence would provide a wealth of information. But what kind of information, biological or physical? Biological, Collins tells us.

What is the difference? Biological information is information about the functions of systems. Physical information is information about physical composition and structure. Knowledge of physical composition and structure is of little use in determining function. For example, knowing that what British speakers call a "rubber" is composed of that material is of little help in learning its use. The American term, "eraser," identifies the same object by its function. It is such functional identifications that biologists seek first. Uncovering structure is a later goal of inquiry that requires prior functional or biological information. But characterizing the primary sequence of the human genome is a paradigmatically structural inquiry, not a functional one. By itself sequence data can provide no information about functional units like genes, or even homologies – similarities of primary sequences – among them.

Despite Collins's claim, the sequences so far obtained for simpler organisms do not demonstrate the existence of hitherto unsuspected genes, because sequence data cannot do this. In genomes as complicated as the mammalian one, sequence data can only help localize genes – functional units – whose existence has already been established by genetic techniques – i.e. breeding experiments or pedigree studies in the lineage of interest or some suitable model organism. Establishing sequence homologies between known genes of model organisms and parts of the DNA of humans is of the highest biological importance. But first, we must have identified the gene of the model system, then localized and sequenced it, and finally hybridize it with human genomic DNA. The result will be the identification of a human gene, which can then be sequenced! Note, prior sequencing of the whole human genome need play no part in this process, and cannot help divide up the human genome into genes. And except for genes which are physically unique to our species, of which there is no reason to think there are any, knowing the whole sequence of the human genome is unnecessary and insufficient for the identification of any genes. It certainly cannot tell us how many genes there are! Nevertheless, Collins and Galas's statement unequivocally commits the project to providing the whole sequence long before we have the functional information that might make it worth having.

In an article based on an interview with Director Collins, *Science* reporter Leslie Roberts wrote:

> Despite the slow progress, there is little sentiment for abandoning the goal of all-out sequencing.... But some thought is being given to a short cut called one-pass sequencing. The original plan calls for sequencing the whole genome several times to ensure an error rate of 0.001%. "Suppose we try one pass coverage with 1% error rate but it only costs one-tenth as much?" asks Collins. The idea, then, would be to return to the really *interesting* regions and sequence them again. (Roberts, 1993, p. 21)

The question raised by this claim is what the interesting regions are, how they are identified, and what can we do once we have identified them. But the answers to none of these questions will emerge from a complete sequence of the human genome. And that is what is wrong with Watson's claim quoted above as well. The interesting regions are the ones whose gene-products are implicated in the production of proteins in the ribosomes, the regions of DNA which produce messenger RNA for enzymes and other proteins, the regions which produce enzymes that control the gene's expression of other enzyme- and protein-producing parts of the DNA, or which control the controllers, etc. How are these genes to be identified? Not from "the bottom up," because we do not know *ab initio* about which DNA sequences do any of these things; rather, we must work from gene-products back to DNA sequences.

It is true that once we know where a functional gene's sequence lies on the chromosome and can isolate it, the sequence can tell us a great deal about the gene-product. This is because the genetic code has been broken and we

232

know which amino acid molecule – the building block for proteins – is coded by which sequence of three DNA bases – the so-called codon. Instead of analyzing proteins directly, the gene sequence can tell us more about the protein it codes for, and tell us faster, because we have technical means of breaking down the DNA to read off its sequence, and we do not have equally powerful techniques for breaking down the protein and reading off its sequence of amino acids. It is the advent of these means for sequencing DNA which made nucleic acids the locus for biotechnological research to design new pharmaceuticals, and whetted the molecular geneticist's appetite for the HGP.

To establish the existence of a gene (or a mutation within a gene) functionally requires identifying its product – the phenotype which it codes for and which assorts in a family of organisms in accordance with well-known regularities of population genetics. Suppose the phenotype is an abnormality, as it often is in current research, which results from a biochemical defect in the gene-product. In this case, we may be able to make very great therapeutic advances without knowing much about the full DNA sequence, where it is, how long it is, how often it is repeated, etc. All we need to do is isolate the messenger RNA (mRNA) for the particular gene-product, usually expressed most heavily in normal organs and tissues whose defective function we seek to understand. Thus, the brain will be richer in gene-products which effect neural processes than the liver; and the richest harvest of mRNAs from the liver will be for gene-products involved in storage of glycogen, and filtering of hemoglobin. Having identified high concentrations of the mRNA distinctive to the tissue, without proceeding back to the gene, we can employ well-known methods[2] to make synthetic genes and large numbers of copies of them. Once inserted in the right vectors, these synthetic genes can produce enough product to provide drugs which will treat hereditary disorders.

It has not escaped the attention of researchers that being able to produce these synthetic genes – complementary or cDNAs – from mRNAs is a far more attractive route to understanding gene expression and gene regulation than sequencing the whole genome. Indeed this realization is at the basis of a potentially lucrative research program in pharmacology. After all, by mixing, say, radioactively labeled cDNAs with the genome, and observing where the radioactive cDNA molecules bind to the genome or "hybridize," molecular geneticists can zero in on the 5 to 10% of the DNA that codes for gene products – i.e. the genes – without wasting any time sequencing junk. Accordingly, few researchers are asking for government subsidies to produce cDNAs; instead they are attempting to patent them.

Of course it is true that once we have identified a gene through its function – sometimes the immediate gene-product, sometimes an observable phenotypic trait – we can begin to try to localize the gene to a chromosome. What is needed is to identify a detectable abnormality – a landmark – in the chromosome which co-varies systematically with the product or phenotype. Once we have done this we can zero in on the locus on the chromosome and sequence the DNA to find out exactly what the genetic cause is. For this reason it is important to produce physical maps of the genome at a fairly high degree of resolution. However, such a physical map is very far from the full sequence. It is composed of tens of thousands of sequence tagging sites – each a few thousand base pairs long not repeated elsewhere in the genome, of which 90 to 95% will be unrelated to any functional domain of the DNA. Constructing such a map is an integral part of the HGP. Indeed, many scientists and commentators will say it is the most important part of the project. But even a physical map of landmarks is without interest in the absence of pedigree studies that can reveal when it is worth analyzing chromosomes for distinctive markers that will physically localize genes to DNA sequences. To be useful physical maps and DNA sequences need to be preceded by genetic research that can narrow down the focus of sequencing from the whole genome to the functional units.

Those who identify physical mapping as the central goal of the HGP implicitly demote sequencing the whole human genome to a derivative status. Far from being derivative, it should be no part of the project's expected goals at all.

3   PRIVATE CHOICE AND PUBLIC GOODS

Despite the problems of top-down science, despite its doubtful payoff in worthwhile scientific information when compared to other problems in molecular biology at which we could throw money, few in the molecular genetics community dissent from support for the HGP's objective of sequencing the entire human genome. From the perspective of rational choice theory this should be no surprise. Given the aims and objectives of scientists, the biotech companies, given the patent laws, and given the nature of the human genome, the incentive molecular biologists have to secure a government-supported project to sequence the human genome is very strong.

The entire sequence of the human genome is a piece of information. And to produce this information, a good deal of further information needs to be provided. If the allegory of the phone books is apt, given the present state of knowledge, having the entire human genome "on disk" is of little informational value. However, there is no doubt that the research program of the HGP will generate a large number of technological breakthroughs necessary for producing the sequence: research methods, computer programs – informatics, reagents, assays, pieces of machinery for automated sequencing, etc. The HGP requires the joint problem-solving skills of biochemists, physicists, engineers, mathematicians and computer scientists. The real value of the breakthroughs required to sequence the whole genome is that as "spin-offs" they are crucial to answering important currently pressing research questions that the scientists and commercial biotechnology firms have. These questions are not about the sequence of the human genome as a whole, but of course about functional units of it which have scientific and commercial interest. It may well turn out of course that for reasons no one can now identify, having the entire sequence of the human genome will be valuable information. But the supposition that it will be is a very expensive gamble. Why then the strong pressure from the molecular biological community to establish and maintain the HGP?

We can shed some considerable light on why leading figures in molecular genetics have committed us to the HGP, and perhaps draw some policy-relevant normative conclusions by applying a little of the economics of information to the project.

The first thing to know about information generally is that it is unlike most other commodities. Other things being equal, economically valuable information always benefits larger scale producers over smaller ones: since the costs of discovery are the same, but can be spread over a larger production run by larger producers. This means that larger concerns can outbid smaller ones for information, and that concerns with exclusive rights to some valuable information will grow larger than others. The result will be a loss of general welfare well understood in the economics of monopoly and monopolistic competition.

The second thing of importance about information is its inappropriability. An individual who has acquired some information cannot lose it by selling or giving away to others a non-exclusive right to use it. Moreover, information acquired at great cost in research, or purchased for non-exclusive use, can be sold very cheaply, or even given away. Consider what happens to a piece of software newly introduced in an office. The market price of a token of information generally is much smaller than the cost of research to develop the first token. The information cannot be fully appropriated by a purchaser since the seller can sell to others. Nor can a seller recapture its full commercial value in the price charged for selling a token of the information, because this value includes returns to purchasers who resell. Thus, the production of new information is always below the optimum level, *ceteris paribus*. On the other hand, if firms kept information secret, did not sell it, there would be overinvestment in research and development, as firms duplicated one another's findings. Thus, according to established economic analyses, a competitive economy always underinvests in research and development. This is a form of market failure. (See Arrow, 1984.)

Because it is inappropriable, and not provided at optimum levels, information has an economic role rather like that of a public good. An economically rational individual will invest resources in the discovery of new information up to the level of the marginal expected value of the information to the discoverer. But the expected value of the new information to all researchers is the sum of the marginal expected value to each of them, and will be larger than the expected value to the individual discoverer or inventor alone. Unless some part of the marginal value to other potential users can be captured by the discoverer or inventor, he will lack the purely economic incentive to invest in research up to anywhere near the point at which the costs of research equal the benefits to the whole economy of the research. Whence underinvestment because of the scope for "free-riding."

Solutions to problems of market failure in the provision of public or other inappropriable goods usually involve governmental coercion. In the case of information, such coercion comes in the form of patent protection. The solution to the problem of undersupply is the establishment of governmentally enforced copyrights or patent rights in the discovery of the information. Patent and copyright protection enables the first discoverer to secure some of the marginal value that other users can produce by implementing the discoverer's information. Thus it encourages individual investment at levels closer to the optimal level of investment for the whole economy. Patent protection is not a perfect solution to the underproduction problem; it cannot shift the market to a welfare-optimum equilibrium. No invention has a perfect substitute, and so an exclusive right to sell the invention allows its owners to charge a noncompetitive price for their information. This reduces consumption of information by raising its cost. Thus, patents and copyrights alleviate the underproduction problem while incurring an under-utilization problem. (See Herschleifer and Ryley, 1992.)

Now, in the case of much biotechnological research, the resulting equilibrium does lead to investment in research. Witness the development of new products, the explosion of new firms, and their attraction of venture capitalism. But in the case of the DNA sequence of *Homo sapiens*, patent protection is unlikely to have the same mitigating effects. Patents are unlikely to result either for the sequence or the spin-offs being produced at anything near optimum levels.

To begin with, there are reasons to think that neither the whole sequence nor large portions of it are open to patent protection. U.S. patent law restricts

patents to new or useful processes, machines, manufactures, or composi-
tions of matter, or useful improvements of them, including genetically en-
gineered living organisms. (Cf. Eisenberg, 1992, p. 227; Sgaramella, 1993,
pp. 299–302.) But it excludes as unpatentable those scientific discoveries
that cannot be immediately used for any human purpose. Absence of immediate utility would presumably exclude both the primary sequence of
base pairs and most physical maps of the human chromosome until discovery of the functional units their landmarks are correlated with. Patents are
also unavailable for so-called "obvious" extensions or applications of prior
information. So, attempts to patent large numbers of cDNAs produced by
means well understood to all researchers have been challenged as failing the
"non-obviousness" test. Since these cDNAs are far more immediately useful
than either the whole sequence, large parts of it, or a physical map of the
whole chromosome, the likelihood of patent protection for any of these latter
seems low.

However, usefulness and non-obviousness are tests which some of the
HGP's technological spin-offs may satisfy. Technology useful for mapping
and sequencing alone may not be any more patentable than the sequences are
either because of non-usefulness or obviousness. Thus, patents and copyrights
alleviate the underproduction problem while incurring an underutilization
problem. (See Herschleifer and Ryley, 1992.) Even if they were patentable,
these innovations would probably be restricted in their foreseeable use to
the HGP labs alone, and so not be lucrative enough to net returns to their
discoverers. However, some of the spin-off technologies will be of great
value in producing useful things like gene-products with great pharmaceu-
tical value. Thus, no single researcher or group has an economically mea-
surable incentive to provide the whole sequence, because there is no patent
protection available for it, nor have researchers material incentives to produce
technologies that do nothing but sequence. Moreover, beyond the most obvious improvements in informatics and sequencing technologies, researchers
cannot tell when spin-offs will turn out to be of great value, and will be
patentable.

Eventually, the sequence will have value, once we have a great deal more
functional information. But no one can predict how much value it will have,
scientific or economic. Similarly, the spin-offs are sure to be harnessed to technological breakthroughs. But the probable time required for breakthroughs
to be harnessed profitably is long, and the probabilities that any particular lab
will secure returns from these breakthroughs is very small. Compare the tran-
sistor or the laser: the former was at first expected only to help produce better
hearing aids; the latter was invented at Bell Labs but was almost not patented,

since it had no evident application in telephone technology, and would not until the advent of fiber optics twenty-five years later. No one anticipated its role in the household CD player.

No lab has an economic incentive to invest in the production of the sequence or the spin-offs alone, nor a non-negligible rational expectation of inventing patentable and lucrative spin-offs.

How should an (economically) rational molecular genetics laboratory respond to this degree of uncertainty about information which may or may not be patentable, the well-known consequences in underinvestment in research and development, and the potential public goods effects of the sequencing project for the molecular genetics community? Let us raise the question for labs, and not for individual researchers, both because the unit of research nowadays is the lab, and because in the present context, their roles are rather like those of individual business firms – in fact, some of the participants in the HGP are private companies. How a rational molecular genetics lab might respond to this state of affairs differs, depending on whether the lab is already a relatively large one or not.

The fact that sequence and spin-off information is a public good for molecular genetics labs is no incentive to the individual researcher to provide it; it is an incentive to free-ride on the willingness of others to provide the good. Could researchers in molecular genetics enter into an enforceable agreement among themselves to produce the sequence and the spin-offs? They could do so, provided they were willing to work together to secure a coercive agent to enforce the agreement, and to provide the resources to develop the technology to determine the sequences.

This last requirement is the most difficult one of course. Not only do individual labs not have control over sufficient disposable research funds to pool together, but if they did, each laboratory would have an incentive to understate its resources. However, suppose they mutually agreed to propose that the government provide these resources, subject to each principal investigator's surrendering the patent right to information that might be generated. In return for this subvention, the larger labs surrender their right to patent sequences and some spin-offs.

If the analysis of patent rights sketched above is correct, the expected value of this right is very low: the sequence-information that is predictably forthcoming cannot be patented and the spin-off information which might be patentable can't be predicted at all for any given lab. The expected value of patent rights is thus equally low for small labs and large ones. Accordingly, large labs will surrender patent rights "cheaply," and small labs will not complain, even though they have not been made parties to the contract, and secure no governmental support from it. Note that federal support of the HGP does not prohibit the patenting by researchers of useful, non-obvious discoveries made while supported by HGP funds.

So, by offering the sequence of the human genome as a non-patentable public good in exchange for the funds to carry out HGP research, the labs give up little in return for a great deal. And the larger the lab, the greater economies of scale it will be able to apply for any unit of foreseeable spin-off information. Corporations pursuing research in biotechnology will happily support this strategy, since their immediate commercial interests are limited to foreseeable spin-off technology, and the expected value of this technology for each of them is sufficiently similar so that none will have an incentive to oppose a no-patentability policy.

The HGP is a good deal for the molecular genetics community. Or rather, it is a good deal for those participants who can secure recognition as a HGP research center, or association with a center. This will include the larger labs in universities, and of course the national laboratories operated for the Department of Energy by leading universities. And to the extent that the allocation of funds for the HGP does not reduce support for other molecular biological research, smaller laboratories which have no stake in the HPG will have no grounds for complaint, and some interest in acquiescing: they may profit from the spin-offs – reagents, assays, informatics, etc. And the small labs have incentives not to criticize the programs of established figures in the field who have advocated the HGP in public debate.

But is the trade-off – three billion dollars in exchange for the sequence and the spin-offs that may not be patentable anyway – worth it to the governments and citizens whose taxes support the research? Is the exchange between the large labs and the government equitable? Beyond equity, does it provide incentives to pursue the most pressing questions, the most promising lines of inquiry, the highest quality research, in molecular biology? Affirmative answers to these questions are doubtful. The multiplication of small, medium, and large biotechnology firms – none of them blindly sequencing the human genome, all of them seeking the sequences that code for gene-products – suggests that the HGP is not the most promising line of inquiry for biomedical discoveries; interest in other sequences, like those of yeast, *E. coli*, *C. elegans*, the mouse, seems very great, but little of it is devoted to sequencing for its own sake the entire genomes of these creatures.

If the government's intention in providing support for the HGP is its expected payoff for health care, then the exchange cannot be a fair one. For were the expected payoff high enough, the costs would have been internalized by the molecular genetics community, especially the biotech firms. If the government's intention in providing support for the HGP is simply to support the best research being done, then surely the bottom-up system which has worked well in the National Science Foundation, the National Institute of Health, and the traditional grant-award processes, would more fully insure high-quality work on the most pressing problems. Of course, if sequencing the human genome were a scientifically valuable immediate objective, given current knowledge, then the dimensions of the task, the efficiencies of the division of labor, the opportunity to spread capital costs widely, and the need to avoid duplication, would make top-down funding and direction sensible. But the antecedent of this conditional does not obtain. The whole sequence has no immediate value, and research into its functional domains can best be carried out under the auspices of bottom-up research.

The HGP thus raises some profound questions for the political economy of large-scale scientific research and governmental support. Exponents of government support for scientific research justify the support by appeal to its eventual instrumental value to the whole society and its intrinsic value in expanding understanding of the way the world works. Even those who might doubt that such objectives are within the proper sphere of government can accept another argument for the public support of science. The regulatory burden – animal care and human subjects oversight committees, toxic waste prohibitions, fair employment practices, public disclosure requirements, drug-free workplace, conflict of interest, anti-lobbying, and other certifications – has so vastly increased the non-experimental costs of scientific research, in the United States at least, that its government is obliged to fund these intrusive mandates, in science and elsewhere.

But everything the theory of public choice tells us about the behavior of individuals and institutions suggests that the government must structure and administer the support of scientific research to maximize decentralization in decisions about the use of its funding. Only decentralizing the decision making about research subjects and strategies to the lowest level possible will harness most effectively the distributed knowledge of the scientific disciplines. Like other human institutions, modern science provides individuals with strong incentives to centralize decision making for their own direct interests or the interests of institutions that indirectly benefit them. The HGP seems a transparent example of this tendency. The traditional system of providing support on the basis of peer reviews of individual proposals is by no means free from the defects of private choice in the public sphere. But it is manifestly freer from these defects than HGP's state central planning mold.

NOTES

Reprinted with permission from *Social Philosophy and Policy* 13:2 (Summer 1996): 1–17.

1. Molecular biologists are sensitive to the fact that calling 95% of the human genome "junk DNA" undercuts the rationale for sequencing the whole genome. As a result, some are suggesting that the scientific community was overhasty in coming to the unanimous conclusion that DNA sequences with no known possible function are "nonsense."
2. In particular, two discoveries make this procedure possible: a viral enzyme, reverse transcriptase, has been isolated, which will build a complementary DNA chain on to a single-stranded DNA or RNA template. Polymerase chain reactions, chemical processes which can be fully automated, enable another enzyme to build up vast numbers of copies of a sequence of DNA bases quickly and cheaply.

Bibliography

Alchian, Armen. 1950. "Uncertainty, Evolution and Economic Theory," *Journal of Political Economy* 58: 211–221.

Alexander, R. 1979. *Darwinism and Human Affairs*, Seattle, University of Washington Press.

Arrow, Kenneth. 1984. *The Economics of Information*, Cambridge, Mass., Belknap Press.

Barkow, J., Tooby, J., and Cosmides, L. 1992. *The Adapted Mind*, Oxford, Oxford University Press.

Bateson, P. 1978. "Review of *The Selfish Gene*," *Animal Behaviour* 26: 316–318.

Beatty, J. 1981. "What's Wrong with the Received View of Evolutionary Theory?" in P. Asquith and R. Giere (eds.), *PSA 1980*, v. 2.

Bennett, Jonathan, 1976. *Linguistic Behavior*, Cambridge University Press; second edition, Hackett, 1990.

Bonjour, L. 1995. "Against Naturalistic Epistemology," *Midwest Studies in Philosophy* 19: 283–300.

Boyd, R. 1973. "Realism, Underdetermination, and a Causal Theory of Evidence," *Noûs* 7: 1–12.

Boyd, R. 1980. "Scientific Realism and Naturalistic Epistemology," in P. Asquith and R. Giere (eds.), *PSA 1980*, v. 2, pp. 613–662.

Boyd, R. 1985. "How to Be a Moral Realist," in J. Sayre-McCord (ed.), *Moral Realism*, Ithaca, N.Y., Cornell University Press.

Boyd, R. 1989. "Realism, Approximate Truth, and Philosophical Method," in W. Savage, ed., *Scientific Theories*, Minneapolis, University of Minnesota Press.

Burge, T. 1979. "Individualism and the Mental," *Midwest Studies* 4: 73–121.

Cherniak, C. 1986. *Minimal Rationality*, Cambridge, Mass., MIT Press.

Chisholm, R. 1977. *Theory of Knowledge*, Englewood Cliffs, N.G., Prentice-Hall.

Cohen, M. 1977. *Food Crisis in Prehistory*, New Haven, Conn., Yale University Press.

Cohen, L. J. 1981. "Can Human Irrationality Be Experimentally Demonstrated?" *Behavioral and Brain Sciences* 4: 317–331.

Collins, F., and Galas, D. 1993. "A New Five Year Plan for the US Human Genome Project," *Science* 262.

Cummins, R. 1975. "Functional Analysis," *Journal of Philosophy* 72: 741–765; reprinted in E. Sober, *Conceptual Issues in Evolutionary Biology*, Cambridge, Mass., MIT Press, 1984.

Darwin, Charles. 1859. *On the Origin of Species*, London, John Murray.

Davidson, D. 1984. *Inquiries into Truth and Interpretation*, Oxford, Oxford University Press.

Dawkins, Richard. 1976. *The Selfish Gene*. Oxford, Oxford University Press.

Dennett, D. 1969. *Content and Consciousness*, London, Routledge.

Dennett, D. 1991. "Real Patterns," *Journal of Philosophy* 88: 27–51.

Dennett, D. 1995. *Darwin's Dangerous Idea*, New York, Harper and Row.

Dobzhansky, T. 1973. "Nothing in Biology Makes Sense Except in the Light of Evolution," *American Biology Teacher* 35: 125–129.

Dretske, F. 1979. *Knowledge and the Flow of Information*, Cambridge, Mass., MIT Press.

Eisenberg, R. 1992. "Patent Rights in the Human Genome Project," in G. Annas and S. Elias (eds.), *Gene Mapping*, New York, Oxford University Press.

Fine, A. 1986. *The Shaky Game*, Chicago, University of Chicago Press.

Fischer, J. M. 1991. "Thoughts on the Trolley Problem," in Fischer and Ravizza (eds.), *Problems and Principles*, New York, Holt, Rinehart, and Winston.

Fodor, J. 1981. *Representations*, Cambridge, Mass., MIT Press.

Fodor, J. 1984. "Observation Reconsidered," *Philosophy of Science* 51: 23–42.

Fodor, J. 1990. *A Theory of Content*, Cambridge, Mass., MIT Press.

Fodor, J. 1994. *The Elm and the Expert*, Cambridge, Mass., MIT Press.

Foot, Phillipa. 1967. "The Problem of Abortion and the Doctrine of Double Effect," *Oxford Review* 5.

Friedman, Milton. 1953. "Methodology of Positive Economics," in *Essays in Positive Economics*, Chicago, University of Chicago Press.

Gehring, W., Halder, G., and Callaerts, P. 1995. "Induction of Ectopic Eyes by Targeted Expression of the *Eyeless* Gene in Drosophila," *Science* 267 (1995): 1788–1792.

Giere, R. 1988. *Explaining Science*, Chicago, University of Chicago Press.

Griffiths, P. E., and Gray, R. D. 1994. "Developmental Systems and Evolutionary Explanation," *Journal of Philosophy* 91: 277–304.

Griffiths, P. E., and Gray, R. D. 1997. "Replicator II – Judgement Day," *Biology and Philosophy* 12, 471–490.

Hardin, C. L. 1988. *Color for Philosophers*, Indianapolis, Hackett.

Hardin, C. L. 1992. "The Virtues of Illusion," *Philosophical Studies* 68: 371–382.

Harner, M. 1970. "Population Pressure and the Social Evolution of Agriculturalists," *Southwest Journal of Anthropology* 26: 67–86.

Harner, M. 1977. "Ecological Basis for Aztec Sacrifice," *American Ethologist* 4: 117–135.

Harris, M. 1979. *Cultural Materialism*, New York, HarperCollins.

Herschleifer, J., and Ryley, J. 1992. *The Analytics of Uncertainty and Information*, Cambridge, Cambridge University Press.

Horder, T. J. 1989. "Syllabus for an Embryological Synthesis," in D. B. Wake and G. Roth (eds.), *Complex Organizational Functions: Integration and Evolution in Vertibrates*, New York, John Wiley.

Hull, David. 1989. *Science as a Process*, Chicago, University of Chicago Press.

Hume, D. 1888 [1737]. *A Treatise of Human Nature*, Oxford, Clarendon Press.

Bibliography

Israel, D. 1987. *The Role of Propositional Objects of Belief in Action*, Stanford, CLSI Monographs, pp. 87–72.

Kahneman, D., Slovic, P., and Tversky, A. 1982. *Judgment Under Uncertainty: Heuristics and Biases*, Cambridge, Cambridge University Press.

Kim, J. 1993. *Supervenience and Mind*, Cambridge, Cambridge University Press.

Kitcher, P. 1984. "1953 and All That: A Tale of Two Sciences," *Philosophical Review* 93, 335–373.

Kitcher, Philip. 1989. "Explanatory Unification and the Causal Structure of the World," in Kitcher and Salmon (eds.), *Scientific Explanation: Minnesota Studies in Philosophy of Science*, v. 13, pp. 410–505.

Kitcher, Philip. 1992. "The Naturalist Returns," *Philosophical Review*, 101: 53–114.

Kitcher, Philip. 1993. *The Advancement of Science*, Oxford, Oxford University Press.

Kitcher, P., Sterelny, K., and Waters, K. 1990. "The Illusory Richness of Sober's Monism," *Journal of Philosophy* 87, 158–161.

Kornblith, H. 1994. *Naturalistic Epistemology*, second edition, Cambridge, Mass., MIT Press.

Land, E. 1977. "The Retinex Theory of Color Vision," *Scientific American*, 237.6: 108–128.

Laudan, L. 1977. *Progress and Its Problems*, Berkeley, University of California Press.

Laudan, L. 1981. "The Pseudoscience of Science," *Philosophy of the Social Sciences* 11: 173–198.

Laudan, L. 1984a. *Science and Values*, Berkeley, University of California Press.

Laudan, L. 1984b. "A Confutation of Convergent Realism," *Philosophy of Science* 48: 19–48.

Laudan, L. 1987. "Progress or Rationality: The Prospects for Normative Naturalism," *American Philosophical Quarterly* 24: 19–31.

Laudan, L. 1994a. Comments on Kitcher, APA Pacific Division Annual Conference, unpublished.

Laudan, L. 1994b. "Realist Reveries and Comparativist Realities: Comparativism v. Eliminativism," University of California, Riverside, Annual Philosophical Conference, unpublished.

Lawrence, P. 1992. *The Making of a Fly: The Genetics of Animal Design*, Cambridge, Mass., Blackwell Scientific Publishers.

Leplin, J. 1982. *Scientific Realism*, Berkeley, University of California Press.

Lewis, D. 1974. *Counterfactuals*, Cambridge, Mass., Harvard University Press.

Lewis, D. 1986. *On Plurality of Worlds*, Oxford, Blackwell.

Lewontin, R. 1980. "Theoretical Population Genetics in the Evolutionary Synthesis," in Mayr, E., and Provine, W. (eds.), *The Evolutionary Synthesis*, Cambridge, Mass., Harvard University Press.

Lewontin, R., and Sober, E. 1982. "Artifact, Cause and Genic Selection," *Philosophy of Science* 47: 157–180.

Lloyd, E. 1993. *The Structure and Confirmation of Evolutionary Theory*, Princeton, Princeton University Press.

Lovejoy, A. 1908. "The Thirteen Pragmatisms," *Journal of Philosophy* 5: 1–12.

Mayr, E. 1982. *The Growth of Biological Thought*, Cambridge, Mass., Belknap Press.

McCloskey, Donald. 1985. *The Rhetoric of Economics*. Madison, University of Wisconsin Press.

Millikan, R. 1984. *Language, Thought, and Other Biological Categories*, Cambridge, Mass., MIT Press.

Moore, G. E. 1903. *Principia Ethica*, London, Routledge and Kegan Paul.

Morgan, T. H. 1897. *Wilhelm Roux Archives* 5: 582.

Nagel, E. 1956. *Logic without Metaphysics*, Glencoe, Ill., Free Press.

Nagel, E. 1961. *The Structure of Science*, New York, Harcourt Brace, 1961; reprinted Indianapolis, Hackett, 1986.

Nagel, E. 1977. "Teleology Revisited," *Journal of Philosophy* 74: 261–301.

Nelson, Richard, and Winter, Sidney. 1982. *An Evolutionary Theory of Economic Change*, Cambridge, Mass., Belknap Press.

Papineau, D. 1994. *Philosophical Naturalism*, Oxford, Blackwell.

Petrowski, R., and Rey, G. 1995. "Saving *Ceteris Paribus* from Vacuity," *British Journal for the Philosophy of Science* 46: 81–110.

Quine, W. V. O. 1951. *From a Logical Point of View*, Cambridge, Mass., Harvard University Press.

Quine, W. V. O. 1960. *Word and Object*, Cambridge, Mass., MIT Press.

Quine, W. V. O. 1964. *Ontological Relativity*, New York, Columbia University Press.

Quine, W. V. O. 1969. *Ontological Relativity and Other Essays*, New York, Columbia University Press.

Quine, W. V. O. 1990. *Pursuit of Truth*, Cambridge, Mass., Harvard University Press.

Railton, Peter. 1986. "Moral Realism," *Philosophical Review* 95: 163–207.

Rescher, N. 1977. *Methodological Pragmatism*, Oxford, Blackwell.

Rescher, N. 1998. *Predicting the Future*, University of Pittsburgh Press.

Rifkin, Jeremy. 1984. *Algeny*, New York, Penguin.

Roberts, L. 1993. "Taking Stock of the Genome Project," *Science* 262.

Rosenberg, A. 1985. *The Structure of Biological Science*, Cambridge, Cambridge University Press.

Rosenberg, A. 1986. "Intentional Psychology and Evolutionary Biology, Part I, Part II," *Behaviorism* 14: 15–28, 125–138.

Rosenberg, A. 1994. *Instrumental Biology or the Disunity of Science*, Chicago, University of Chicago Press.

Rosenberg, A., and Martin, R. M. 1979. "The Extensionality of Causal Contexts," *Midwest Studies* 4: 401–408.

Russell, B. 1910. *Philosophical Essays*, London, Longmans.

Sayre-McCord, J. *Moral Realism*, Ithaca, Cornell University Press.

Schaffner, K. 1993. *Discovery and Explanation in Biology and Medicine*, Chicago, University of Chicago Press.

Sgaramella, V. 1993. "Lawyers' Delights and Geneticists' Nightmares: At Forty, the Double Helix Shows Some Wrinkles," *Gene* 135: 299–302.

Simon, H. 1957. *Models of Man*, New York, Wiley.

Smith, John Maynard. 1984. *Evolution and the Theory of Games*, Cambridge, Cambridge University Press.

Sober, E. 1978. "Psychologism," *Journal for the Theory of Social Behavior* 8: 165–191.

Sober, E. 1984. *The Nature of Selection*, Cambridge, Mass., MIT Press, reprinted 1993, University of Chicago Press.

Sober, E. (ed.). 1984. *Conceptual Issues in Evolutionary Biology*, Cambridge, Mass., MIT Press.

Sober, E. 1993. *The Philosophy of Biology*, Boulder, Colo., Westview.

Sober, E., and Wilson, D. 1998. *Do Unto Others*, Cambridge, Harvard University Press.

Sterelny, K., and Kitcher, P. 1988. "The Return of the Gene," *Journal of Philosophy* 85, 339–361.

Sterelny, K., Smith, K. C., and Dickison, M. 1996. "The Extended Replicator," *Biology and Philosophy* 11, 377–403.

Stich, S. 1983. *From Folk Psychology to Cognitive Science*, Cambridge, Mass., MIT Press.

Stich, S. 1990. *Fragmentation of Reason*, Cambridge, Mass., MIT Press.

Sturgeon, N. 1985. "Moral Explanations," in Copp and Zimmerman (eds.), *Morality, Reason, and Truth*, Totowa, N.J., Rowan and Littlefield, pp. 49–78.

Taylor, Charles. 1964. *The Explanation of Behavior*, London, Routledge.

Thompson, P. 1988. *The Structure of Biological Theories*, Albany, SUNY Press.

Thomson, J. J. 1986. "Killing, Letting Die and the Trolley Problem," and "The Trolley Problem," in W. Parent (ed.), *Rights, Restitution and Risk*, Cambridge, Mass., Harvard University Press.

van Fraassen, B. 1979. *The Scientific Image*, Oxford, Oxford University Press.

Waters, K. 1990. "Why the Antireductionist Consensus Won't Survive: The Case of Classical Mendelian Genetics," *PSA 1990*, East Lansing, The Philosophy of Science Association.

Watson, J. B. 1993. "Looking Forward," *Gene* 135: 309–315.

Wilson, E. O. 1978. *On Human Nature*, Cambridge, Mass., Harvard University Press.

Wolpert, L. 1969. "Positional Information and the Spatial Pattern of Cellular Formation," *Journal of Theoretical Biology* 25: 1–47.

Wolpert, L. 1994. "Do We Understand Development?" *Science* 266: 571–572.

Wright, L. 1976. *Teleological Explanation*, Berkeley, University of California Press.

Index

Index

Index

utilities, 165, 166; *v.* disutilities, 167
utility, 160, 166
utility-maximization, 135

value, 120; of capital good, 203
value-neutrality, 138
van Fraassen, B., 10, 15, 26, 27, 31
variation, 105, 121, 125; fortuitous, 174
veil of ignorance, 214
verb, 63
verisimilitude, 20
vicious-circle arguments, 27
Voltaire, 78, 83

Walrasian auctioneer, 186, 207
Walrasian *tâtonnement*, 207, 218; *see also*
    tâtonnement
Wasserman test, 161

Waters, C. K., 96
Watson, J. B., 64, 230
weak generalizations, 59
wealth, 216
Weismann, G., 108
welfare economics, 189
welfare-optimum equilibrium, 236
well-being, 141, 142, 143, 147, 151;
    maximization of, 148–149
Whiggish interpretations, 13, 56
Wieschaus, E., 96
Wilson, D. S., 134
Wilson, E. O., 119
Winter, S., 189–192
Wolpert, L., 76, 79, 89, 91
working posits, 21
World War II, 69
Wright, L., 8, 46, 75, 87